LEARN
&
unLEARN

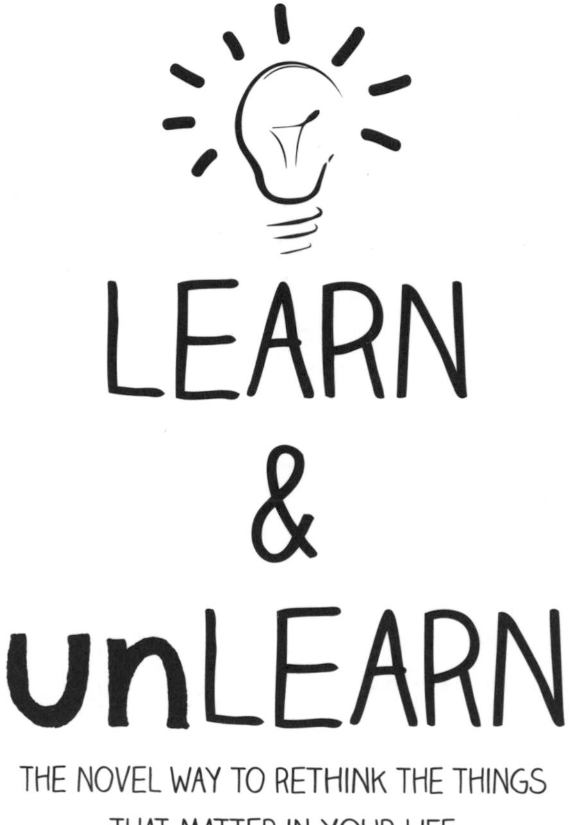

Surendra Verma

This edition published in 2015 by New Holland Publishers Pty Ltd
London • Sydney • Auckland

Unit 009, The Chandlery 50 Westminster Bridge Road London SE1 7QY UK
1/66 Gibbes Street Chatswood NSW 2067 Australia
218 Lake Road Northcote Auckland New Zealand

www.newhollandpublishers.com

A record of this book is held at the British Library and the National Library of Australia.

ISBN 9781742575964

Managing Director: Fiona Schultz
Project Editor: Susie Stevens
Designer: Peter Guo
Production Director: Olga Dementiev
Printer: Toppan Leefung Printing Ltd (China)

10 9 8 7 6 5 4 3 2 1

Keep up with New Holland Publishers on Facebook
www.facebook.com/NewHollandPublishers

For Anuraag Verma
and Rohit Verma

We must unlearn the constellations to see the stars.

Jack Gilbert, 'Tear it down', *Transgressions: Selected Poems*

about the author

Surendra Verma is a science writer, journalist and author based in Melbourne, Australia. His recent popular science books include:

The Mystery of the Tunguska Fireball

Why Aren't They Here?: The Question of Life on Other Worlds

The Cause of Mosquitoes' Sorrow: Beginnings, Blunders and Breakthroughs in Science

The Little Book of Scientific Principles, Theories & Things

The Little Book of Maths Theorems, Theories & Things

The Little Book of Unscientific Propositions, Theories & Things

The Little Book of the Mind: How We Think and Why We Think

+ a children's book

Who Killed T. Rex?: Uncover the mystery of the vanished dinosaurs

Visit www.surendraverma.com for more information on these books.

Contents

Introduction

learn/unlearn/relearn

> A novel is balanced between a few true impressions and the multitude of false ones that make up most of what we call life. It tells us that for every human being there is a diversity of existences, that the single existence is itself an illusion in part, that many existences signify something, tend to something, fulfil something; it promises us meaning, harmony and even justice.
>
> Saul Bellow, Nobel Prize Lecture, 1976

Unlearning is not the same as forgetting. It requires fundamental changes to our mental processes and a relentless willingness to pursue new ways. It's far more difficult than learning as it challenges old values and beliefs. But it opens up new realties and possibilities by unleashing creative energy to experience others and ourselves in absolutely different ways.

A lifetime of learning cannot be erased easily. Therefore, the quest to embrace new knowledge and unlearn old skills and behaviours is a road full of potholes. The old wisdom is that we learn best from our mistakes. Research, however, shows that something more is needed: we must be conscious of our mistakes to harness the benefits of new learning. 'You have to keep in mind the new way of doing things while suppressing the old way,' says Jason Moser, a clinical psychologist at Michigan State University. 'It's a lot of work and hard to overcome first.' The reason is that learning new things leads to more effort, more mistakes and, eventually, a lot of frustration. Once you are consciously aware of the mistake you have made in your learn-unlearn-relearn process, your performance at the new task is likely to improve.

Neurons, or nerve cells, fire electrochemical signals from one to another through special junctions called synapses. Whenever we learn, say, a new word the synapses pass signals along a new circuit. These durable connections among certain neurons form the memory of that word. If certain neural connections are blocked new ones open up. As these new connections continue to be used, they become stronger and more prominent—a phenomenon captured in the oft-repeated phrase 'neurons that fire together, wire together'. This ability of our brains to produce new connections between neurons as a result of new experiences is neuroplasticity.

The adult brain has more than 86 billion neurons and, on average, each fires five to 50 times per second. The signals travel as fast as 435 kilometres per hour through 170,000 kilometres of nerve fibres. Thus our brains, the most complex of machines in the universe, have an enormous capacity to form new connections. These new connections form the basis of learning and memory. In other words, our brains are plastic and they continue to transform, or rewire, throughout our lives with new learning and experiences.

Neuroplasticity, the buzzword of the 21st century, reminds us that we are dynamic beings capable of learning new attitudes, behaviours and ways of living. In the same way, we can harness neuroplasticity for unlearning old things.

This book is not a treatise on neurological and psychological processes involved in learning and unlearning. It simply presents 40 ideas, based on latest researches, which matter in life in our fast-changing world.

If you're wondering what novels have to do with this learning and unlearning, bear with me as I tell you about a novel that has yet to be written:

> She hovers over my desk, her eyes arch as she whispers, 'Write a novel, not another silly science book.' She disappears into the mist but her words alloyed with annoyance linger on in cool air. Following her command, I start working on a novel. A sample paragraph:
>
> 'Fruit and vegetable markets tend to be narrow streets with stalls on both sides; they are not wide avenues lined up with trees and fancy shops selling designer fruits and veggies, at least not yet. This one was no different except the extra charm of hawkers with bicycle-wheel push carts and peddlers on foot spruiking their wares; poor little

girls from hills with faces like pink flowers, their mothers with wrinkled faces and wicker baskets on their heads; housewives with cotton bags in their hands haggling over prices, squeezing fruits to test their firmness and the limits of stall holders' patience while their husbands' look the other way; servants with hessian bags shopping for their miserly mistresses, carefully comparing the highest and the lowest prices of a particular produce so that they can pocket a few coins; young barefoot boys in ragged cotton shirts and pyjamas shivering and begging, and if not begging working at stalls for a pittance; teenage stall workers picking produce from large wicker baskets, putting them on their hand scales, weighing them and throwing them into buyers' bags, all happening so quickly making buyers worried about short-changing; stall holders touting, haggling, peeling leaves and throwing them on the street for the two cows, one lazing at each end of the street, who preside over this two-hundred-yard long spectacle and put on their own show when they stand up and start walking slowly towards the other end of the street cleaning it like an efficient street cleaner, once they reach the middle of the street, they eye each other, ruminate for a while, pee on the street for the scientifically observed duration of exactly twenty-one seconds and then turn back towards their usual lazing spots; a town council cleaner occasionally visiting the market, dipping a finger in the pee and touching it on his forehead as a sign of respect for the cows' holiness, and praying with bowed head and folded hands in front of the cows to thank them for doing his job so well; and less occasionally a temple elephant painted with pigments and adorned with bells and necklaces, and his mahout on foot, both with impeccable street manners, walking through the street, Hindu stall holders rushing with their best banana or two to feed the elephant for blessings for brisk trade from Lord Ganesha, the elephant-headed god, while Muslim stall holders looking on with smiles of scepticism, and tiny toddlers in their mothers' arms screaming with pure joy without any trace of belief or unbelief on their innocent faces radiating in the warmth of the weak wintery sunlight.'

As a science writer, I realise, it's much easier for me to describe imaginary scenes; it's no different from describing a scientific phenomenon. But a science writer's style of writing turns characters into cartoonish caricatures. From this short creative writing exercise I *learn* that I have to *unlearn* many things before I can write a novel. When we want to do something new we must first unlearn old ways of doing things. This idea has led me to write this 'silly science book'. I have turned to novels (a multitude of true and false impressions that make up most of what we call life, as Bellow says) to make up the list of 'things that matter in life' to include in this book. Though the ideas of stories are drawn from novels, they are based on latest researches in science. However, reading them requires absolutely no background in science.

The book provides up-to-date information on various topics of general interest. Please do consult a health professional before you act on any health and diet information in this book.

Learn, unlearn, relearn and wonder!

Chapter 1

From print to screen, do words scream?

learn reading speed and comprehension is consistent for expert readers whether they read on paper or screen, but children may find digital devices distracting/*unlearn* digital medium is essential for children's development (we still do not know the ultimate effects of an immersion in digital medium on young developing brain)

> Langdon returned his focus to the iPhone, and within seconds was able to pull up a link to a digital offering of *The Divine Comedy*—freely accessible because it was in the public domain. When the page opened precisely to Canto 25, he had to admit he was impressed with the technology. *I've got to stop being such a snob about leather-bound books*, he reminded himself. *E-books do have their moments.*
>
> Dan Brown, *Inferno*

The Irish playwright George Bernard Shaw, who won the Nobel Prize in Literature in 1925, once came across one of his books in a second-hand bookshop, inscribed *To *** with esteem, George Bernard Shaw*. He bought the book and sent it back to *** after adding the line, *With renewed esteem, George Bernard Shaw*. If doomsayers are right and e-books do kill the printed books, authors like me can never dream of signing their books 'with renewed esteem' or even 'with esteem', and when visiting a bookshop will also miss the surreptitious pleasure of finding a lone copy and rescuing it from obscurity by placing it on the shelf in a more prominent position.

Forget authors, what about readers?

Some studies suggest that on-screen reading is measurably slower than reading on paper, and it also disadvantages your ability to remember what you have read long-term. At the same time, many studies have found no difference between reading speed and comprehension between paper and screen. On balance, when reading long texts paper still has advantage over screens.

Memory is helped by spatial context. You will find it easier to remember what you are reading on this page if you try to associate it in your mind with the place where you are reading it, such as your bedroom, your office or the train. Similarly, remembering irrelevant facts such as whether this page is at the beginning, middle or end of this book—or a certain fact you have read is at the top or bottom of the page or near a graphic—can help anchor the material in your mind. This oldest known mnemonic trick is called the method of loci: associating information you want to remember with places you already know such as your house. When you want to recall the information, simply visualise yourself wandering through the house, recalling the items as you go through it room by room in your mind.

We're still a long way away from e-books that would allow readers to visualise a coherent mental map of the text as in printed books: an open book's many corners, the thickness of read and unread pages, ability to quickly flip the pages, they all add to spatial context that is essential for making memories stronger.

Mark Changizi, an American theoretical neurobiologist, says that the new reading experiences within e-books have lost their spatial navigability, which can be key to remembering information: 'Need to jump to that part of the book where they discussed cliff jumping? You will get no help from local topography, but you can beam yourself directly there via a within-document text search.'

Smaller screens such as that of mobile phones make materials less memorable. This message comes from (definitely not from the mobile phone) of Jakob Nielsen, a web 'usability' expert who worries about the screen size of e-books: 'The bigger the screen, the more people can remember, and smaller, the less they can remember.' The most dramatic example is reading from mobile phone, he says, you lose almost all context.

Without losing any context, Kate Garland, a psychologist at the University Leicester in England, has extensively studied the question of reading on digital media (in her case computer screen) or on paper. Her verdict: when the exact same material is presented in both media, there is no measurable difference in performance. However, she points out two differences that may matter in the long run: (a) printed book readers seem to digest the material more fully, and (b) more repetition is required with computer reading to impart the same information.

Reading has an evolutionary disadvantage. We as a species learned to speak before we learned to read and write. We didn't invent writing until the fourth millennium BC, a recent phenomenon in our long evolutionary history. It is, therefore, most likely that the human brain has not been purposefully designed—or hardwired, as they say these days—for reading. The neural pathways for reading seem to be borrowed from the regions of the brain that are involved in vision, hearing and language. As these areas are distributed throughout the brain, efficient communication between them is essential for proficient reading.

Research shows that reading ability in children seems to be linked to the strength of connections between neurons in the white matter, which contains nerve fibres connecting distant regions of the brain. 'No one really knows the ultimate effects of an immersion in digital medium on the young developing brain,' stresses Maryanne Wolfe, author of *Proust and the*

Squid: The Story and Science of the Reading Brain. 'We do know a great deal, however, about the formation of what we know as the expert reading brain that most of us possess to this point in history.' It's a well-known fact that children are more successful in reading when they spend their pre-school years reading with parents. However, e-books are not substitute for parents.

The first-of-its-kind study by Julia Parish-Morris, an American developmental psychologist, shows that e-books are becoming a behaviourally oriented, slightly coercive parent-child interaction as opposed to parents talking about the story or making comments that relate to pictures or themes in the book to their child's real life in a way that fires up neurons in the child's brain. 'What was happening with the e-books is that reading was not even part of the process, probably because these books literally read the story to the child,' she says. 'So parents are not needed. The books makes commands and tells the child what to do; it encourages them to play games and reads to the child, so parents are essentially replaced by this battery-operated machine.'

Other studies show that reading comprehension is higher when children read printed books. Children spend a fair bit of their e-book engagement time playing bells and whistles embedded in the e-books rather than reading the text. These distractions disrupt their reading fluidity, which compromises comprehension. Multimedia features in some e-books, however, do help enhance comprehension. It may seem that interactive e-books can entertain children all by themselves, but such books require more input from teachers and parents than conventional books.

We may not yet know the ultimate effects of an immersion in the digital medium on the developing brains of children, but researchers from Johannes Gutenberg University in Germany (sure they know a thing or two about reading books as their university bears the name of the founder of the printing press) inform us that older people find e-books easier to read than the printed page.

In their study, which was also the first of its kind, the researchers decided to test whether reading on digital media requires higher cognitive efforts than reading conventional books. They tracked the eye movements and brain activity of a group of younger participants (aged 21–34 years) and a group of older adults (aged 60–77) as they read nine short texts on e-readers, tablet computers and printed pages. Younger participants showed no difference in time and brain activity when reading on different mediums, but older participants spent more time and effort when reading from the printed page. The best results for older people came when they read from backlit e-readers or tablet computers, which provide increased contrast between the text and background. The study results fail to back up the popular notion that digital reading devices are more tiring on the eyes. No excuses now for not buying a backlit Kindle or an iPad for an old aunt.

Whatever the experts say about e-books or printed books, books are, in the words of Stephen King, the king of the horror novel, 'a uniquely portable magic'. The magic of reading never fades out whether you are reading an e-book or a printed book. Continue reading whether you are holding a (pricey) e-reader or a (priceless) copy signed by the author, *With renewed esteem.*

Chapter 2

Bonheur bilingue

learn to be a smart bilingual; learning another language forces the brain to train permanently from the cradle to the grave/*unlearn* teaching children a second language too early impedes 'normal' learning of their mother tongue and can cause confusion

> 'I may not have been great at law but a further language can be decisive. You know what the poets say, I expect? ... "To possess another language is to possess another soul." A great king wrote that, sir, Charles the Fifth. My father never forgot a quotation, I'll say that for him, though the funny thing is he couldn't speak a damn thing but English.'
>
> – Ricki Tarr to George Smiley
> John le Carré, *Tinker Tailor Soldier Spy*

When John Pentland Mahaffy, celebrated Irish polymath and wit, met Queen Ena of Spain in the 1930s, he entertained her with the well-known division between European languages: 'French to address a friend, Italian to make love to a mistress, Spanish to speak to God and German to give orders to a dog.' If Mahaffy was living in ancient Greece and would have told them this joke, they would have thumbed their noses at him. To ancient Greeks, other languages were gibberish; ironically, nowadays when we say 'it's all Greek to me', we mean we do not understand it at all. 'Greek' meaning 'unintelligible language or gibberish' comes from Shakespeare's *Julius Caesar*; and it reflects the deep-rooted attachment that many of us feel towards our mother tongue. The noted American linguist Uriel Weinrich called it 'language loyalty'.

It may or may not be 'language loyalty', but until recently it was believed that teaching children a second language too early might impede 'normal' learning of their mother tongue. This belief is based on the assumption that the brain has limited learning resources and two languages compete for resources. The belief is reflected in the contemporary education practice, which tends to offer formal schooling in a second language in later school years, not in the developmentally crucial toddler years of learning. Another myth that still perpetrates is that knowledge acquired in one language is not accessible in another language. Everyday experience says something different: if you learn the basic principle of addition in English, you are able to apply this skill to French numbers when you learn French.

Neuroscience has now completely rejected the myth that the brain is set for one language only: learning a second language not only boosts children's brains during infancy, it also protects against decline in brainpower in old people. Brain imaging of monolinguals and bilinguals shows that both process their individual languages in a fundamentally similar way: monolinguals (in one language) and bilinguals (in both languages) show increased activity in language processing areas of the brain. The one fascinating difference is that bilinguals appear to recruit more of the neurons available for language processing than monolinguals. This provides a fascinating insight into the language processing potential not used in monolingual brains.

Laura-Ann Petitto, an American cognitive neuroscientist who is a leading researcher in the new discipline of neuroeducation, says even bilingual parents often opt to 'hold back' one of the family's two languages in their child's early life. 'They believe that it may be better to establish one language firmly before exposing their child to the family's other language so as to avoid confusing the child,' she says. They also worry that earlier bilingual exposure may put their child 'in danger of never being as competent in either of two languages as monolingual children are in one'.

The silent and portable functional near-infrared spectroscopy brain imaging monitors allow researchers to study the brains of babies as they sit on their parents' laps, making this new technique more suitable for studying young children than fMRI (functional magnetic resonance imaging). Pettito's first time use of this new imaging technique to look into the developing brains of bilingual as compared to monolingual children has rejected unequivocally the myth that exposure to two or more languages 'too early' can cause developmental language delay and confusion. Her research supports the idea that bilingualism can invigorate rather than hinder a child's development. It also rejects the flip side of this myth—later exposure is better.

Hungarian psychologist Agnes Melinda Kovacs and her Spanish colleague Jacques Mehler have opened up another fascinating window to the bilingual brain by studying 'crib bilinguals': young bilinguals are more flexible learners. Although infants in bilingual households have to learn roughly twice as much about language as their monolingual peers, the speed of learning is nearly the same for both. It seems that, far from being confused, infants in bilingual households develop superior mental skills or 'executive functions' which play a critical part in complex social behaviour (executive function is our ability to control our thoughts and actions in order to respond suitably to our environment, as opposed to other brain functions dedicated to single tasks such as moving a finger or processing a sound).

Other studies show that executive functions are active not only in bilingual children but also in adult bilinguals. A bilingual person's both language systems are always active and competing, that person uses executive functions every time she or he speaks or listens. This constant practice not only strengthens the prefrontal cortex right behind the forehead, the decision-making region of the brain, but also associated brain regions.

Bilinguals also excel on tasks that require dealing with conflicting information. The brain can perform automated processes quickly and unconsciously; but when it has to process information for more than one task simultaneously, it manages its limited attention resources by inhibiting or stopping one response in order to say or do something else. Bilingual people often perform better than monolinguals on the classic Stroop test (naming aloud the colours of words printed in incompatible ink colour; for example, word 'blue' printed in red ink): everyone takes an additional fraction of a second to accomplish than if both the word and colour are the same. But the lag for bilinguals is measurably shorter; this gives bilinguals lifelong advantage.

Jared Diamond, an American scientist best known for his popular science books, agrees that the clearest difference identified by recent studies involves an advantage that bilinguals have over monolinguals, rather than disadvantage. 'Monolingual people have a special challenge involving executive function,' he says. 'Monolinguals hearing a word need only compare it with their single stock of arbitrary phoneme (sound) and meaning rules, and when uttering a word they draw from that single stock. But multilinguals must keep several stocks separate.'

Obviously, multilinguals have constant unconscious practice in using the executive function system of the brain. This unconscious practice gives older bilingual adults ability to develop new strategies to process language, which helps compensate for age-related decline in cognitive power. A study by Ellen Bialystok, a psychologist at York University in Canada, has found that bilingual people tend to be diagnosed with Alzheimer's disease, the most common form of dementia, four to five years later than monolinguals. She believes that switching between languages strengthens the brain's 'cognitive reserve'—it can be compared to a reserve in a car tank which keeps you going a little longer when you run out of fuel. The idea of 'cognitive reserve' may also explain another finding that older people who speak more than two languages are three times less likely to have memory problems than people who are monolingual.

A study by Thomas Bak of the University of Edinburgh also supports the Canadian finding that those who are fluent in two languages begin to show symptoms of dementia more than four years later than those who only speak a single language. Bak's results were also true for a group of people who were illiterate, suggesting that the benefits of bilingualism are independent of education. His study, conducted on 650 bilingual participants over a six-year period, showed no additional benefits of speaking more than two languages.

In another study Bak looked at the records of 1100 people born in 1936 in and around Edinburgh who were monolingual English speakers at age 11, when they were tested for their cognitive abilities. He tracked down 853 of these people when they were in their early 70s. He found 262 of them had learned to speak a second language and that 65 had learned it after the age of 18. Those who had learned a second language performed better on cognitive tests in their 70s that they did when they were 11. The strongest improvements were seen in general intelligence and reading, indicating that a second language itself is beneficial.

The bilingual brain is constantly suppressing one language and switching between the two. The permanent switching and suppressing offers the best brain training. Older people are encouraged to start new brain-challenging activities such as playing bridge or solving Sudoku puzzles. These activities can engage your brain only for a few hours a day, while bilingual brain is always engaged as it tries to limit interference from the other language to ensure the continued dominance of the intended language.

There are about 6,800 languages in use around the world, of which some 6,500 are spoken languages. Today, more people in the world are bilingual than monolingual. Lazy monolinguals are the so-called illiterates of the 21st century; they are the ones who smirk when people speak the only language they know with a 'funny' accent. If you're not part of the bilingual world, it's time you learned another language to enjoy the bliss of bilingualism. Or, at least, make sure that your children start learning a second language the day they are born. Newborns of bilingual mothers can recognise both languages. Language learning skills start dropping sharply at six years, but they are still far better than yours.

Whether you're a tinker, tailor, solider or spy fluency in another language is indeed like possessing another soul. *Absolument.*

Chapter 3

Forget the pursuit of happiness

learn to have a purpose bigger than yourself/*unlearn* the pursuit of happiness is the most important thing in life

> His real happiness is a ladder from whose top rung he keeps trying to jump still higher, because he knows he should.
>
> – a comment on Harry 'Rabbit' Angstrom's friend Jack Eccles who tries to mend Rabbit's broken marriage
>
> John Updike, *Rabbit, Run*

Key in 'happiness' in Amazon's search box for books and you will be instantly gratified with a selection of tens of thousands of books to lead you on a path of eternal happiness. Or, would they? It's debatable whether reading about mantras and mandates for happiness does help achieve happiness. But the more important question is whether happiness is worth pursuing. In contemporary Western culture personal happiness is one of the most important values in life. In contrast, in some cultures, such as in Japan, the individual pursuit of happiness is perceived as being selfish because it's considered at odds with the good of society.

You will agree that it's good to feel good, but this feel-good feeling is difficult to define and even harder to measure. Happiness is not a static state, it's relative; even the happiest people feel blue at times. Psychologists call this transient state our subjective wellbeing, which is a result of a combination of numerous elements. One of the elements that genuinely lifts the spirit is strong ties to friends and family and commitment to spending time with them. Another is religious faith; it's difficult to find out whether this is because of God or the community aspect or the placebo effect. There is also evidence, scientific evidence, that people who care more about others are happier than more selfish people. The elements that play limited roles in our happiness are: money (once basic needs are met, additional money doesn't upgrade happiness), a good education doesn't necessarily lead to happiness, marriage presents a mixed picture, and youth has no advantages over old age.

No one denies that happiness helps us in the pursuit of our goals, strengthens social bonds and enhances creativity. Like food, happiness is essential for our health and wellbeing. Too much food makes us obese and leads to harmful consequence. Does too much happiness also lead to harmful consequences? In a provocative study American psychologists June Gruber and Iris Mauss and their Israeli colleague Maya Tamir analyse the 'dark side' of happiness and try to answer the questions: is it possible to have too much happiness, to experience happiness in the wrong time, to pursue happiness in the wrong ways, and to experience the wrong type of happiness?

Gruber and colleagues arrive at the following answers:

- Happiness has benefits up to a moderate degree but may be harmful when experienced at an extreme degree. Whereas moderate levels of happiness boost creativity, high levels of positive emotions do not. In fact, happiness 'overdrive' can lead some people to risk-taking behaviours such as binge drinking and drug use.
- Emotions are our responses to circumstances: completing a challenging task fills us with joy, a threat makes us fearful and injustice makes us angry. In a study, participants were motivated to increase their experience of anger when they expected to complete task that required active collaboration, but not those that required collaboration. Participants who were happy performed worse on such tasks than those who were angry, regardless of how they wanted to feel. There are right times to feel happy and right times to feel unhappy.
- The best way to pursue happiness is not to pursue happiness. Studies show that wanting to be happy

> can not only decrease people's wellbeing but also make them lonely. Happiness that causes negative social consequences or that are in conflict with cultural norms may have negative effects. For example, in North American culture happiness tends to be defined in terms of personal pleasure and achievement, while in East Asian culture happiness tend to be defined in terms of social harmony.

A landmark study by American psychologist Barbara Fredrickson and colleagues reveals that happiness has surprising effects on our genes. People who derive happiness from a deep sense of purpose and meaning in life (known as eudaimonic wellbeing, think Pope Francis) show favourable gene-expression profile in their immune cells. They have low levels of inflammatory gene expression and strong expression of antiviral and antibody genes. However, people with high levels of happiness that comes from consuming goods (hedonic wellbeing, think most celebrities) actually show just the opposite. They have an adverse expression profile involving high inflammation and low antiviral and antibody gene expression. The differences in gene expression are independent of factors such age, sex, ethnicity and health.

Happiness derived from contributing to others or to society in a bigger way is different from happiness without meaning that characterises a relatively shallow, self-absorbed or even selfish life, in which things go well, needs and desires are easily satisfied, and difficult and or taxing entanglements are avoided. We know now that these differences in happiness—doing good and feeling good—have very different effects on the human genome, a system of some 21,000 genes that has evolved fundamentally to help humans survive and thrive.

For their study, the researchers drew blood samples from 80 healthy adults who were assessed for hedonic and eudaimonic wellbeing, as well as potentially confounding negative psychological and behavioural factors. They looked at the ways certain genes expressed themselves in each of the participants. While those with eudaimonic wellbeing showed favourable gene-expression profiles in their immune cells and those with hedonic wellbeing showed an adverse gene-expression profile, people with high levels of hedonic wellbeing didn't feel any worse than those with high levels of eudaimonic wellbeing. Both seemed to have the same high levels of positive emotion. However, their genomes were responding very differently even though their emotional states were similarly positive.

'What this study tells us is that doing good and feeling good have very different effects on the human genome, even though they generate similar levels of positive emotion,' the researchers say. 'Apparently, the human genome is much more sensitive to different ways of achieving happiness than are conscious minds.'

The evidence of this study proves that happiness is doing

good and being good, to our bodies at least. It's more than feeling good. Overtly happy people at times appear boring and shallow. A happy life is a life that has a purpose bigger than ourselves. And a 'chemical shortcut' to happiness, hedonic happiness, leads us nowhere.

> 'There are two routes to happiness,' he continues, back at the wheel of the cart. 'Work for it, day after day, like you and I did, or take a chemical shortcut. With the world the way it is, these kids take the shortcut. The long way looks too long.'
> 'Yeah, well, it is long. And then when you've gone the distance, where's the happiness?'
> 'Behind you,' the other man admits.
>
> — Bernie talking to Rabbit at a golf course
> John Updike, *Rabbit at Rest*

Chapter 4

Be free, be open, have no goal

learn to cultivate open-mindedness about the future/*unlearn* not to be obsessive about your goals

> 'When someone is searching,' said Siddhartha, 'then it might easily happen that the only thing his eyes still see is that what he searches for, that he is unable to find anything, to let anything enter his mind, because he always thinks of nothing but the object of his search, because he has a goal, because he is obsessed by the goal. Searching means: having a goal. But finding means: being free, being open, having no goal.'
>
> – Siddhartha to his friend Govinda
> Hermann Hesse, *Siddhartha*

In the 1930s, Indian philosopher J. Krishnamurti warned: 'Do not prepare the mind and heart for the future by shockproof coverings. This is mere self-protection but not intelligence. To be wholly vulnerable is to be wise.'

Let's see what science says about these philosophical musings.

First, what happens in your brain when you are motivated to do a task? Four structures of the midbrain— called the limbic loop—drives our decisions whether or not to act on external and internal stimuli. A small pathway from the limbic system pumps dopamine, the molecule of motivation, into the prefrontal cortex, the 'executive' region of the brain right behind the forehead. When dopamine reaches the frontal cortex we feel good. Dopamine-containing neurons play a vital role in brain networks that govern motivation and a sense of reward and pleasure.

Studies suggest that dopamine is less about reward and pleasure than about drive and motivation. Common sense says that dopamine would be released when we perform a task and receive reward. Experiments on baboons trained to perform a task and receive reward show that dopamine is released just before the baboons perform the task and just before they receive the award, not after. If the reward was not a sure thing but only a possibility, even then the release of dopamine increased substantially. However, when a reward was entirely expected on the basis of the preceding cue, dopamine neurons didn't respond to the reward. Strangely, pleasure declines when we anticipate a reward.

The emerging scientific view is that cultivating open-mindedness about the future leads to positive motivation. Dopamine excels at its task when you set a goal. But setting your mind on a goal and priming it with 'I will' may turn off the tap of your reservoir of intrinsic motivation, especially if you think you would feel guilty or ashamed if you failed.

An intriguing experiment by psychologist Ibrahim Senay of the University of Illinois also confirms the wisdom that setting your mind on a goal may in fact thwart the intended goal. He looked at the problem by exploring self-talk: the voice in your head expressing your options, hopes, fears and so on. Suppose you want to join a gym to go on a regular exercise program. You could do it by talking to yourself in two ways: 'Will I join the gym?' or 'I will join the gym'. Which of these ways of articulating to yourself is the best?

Senay asked a group of volunteers to work on a series of anagrams that required rearranging the word such as 'sauce' to 'cause' or 'when' to 'hewn'. But first he asked them to write either of the two apparently unrelated sentences, 'I will' or 'will I?', 20 times on a sheet of paper. On average, the participants who wrote 'Will I' solved twice as many anagrams as the participants who wrote 'I will'.

Why did asking yourself a question led to better performance than telling yourself? Perhaps an unconscious formation of the question 'will I?' affected motivation. By asking themselves a question, people were more likely to build their own motivation. Senay notes that by unconscious formation of the question 'will I?', the person comes up with their own reasons and gives more thought to what they stand to gain from pursuing a goal or task. He says that self-questioning can be a powerful motivator for change—one that each of us can employ to create a sense of open-mindedness about our current life choices and priorities.

In a follow-up experiment, Senay changed the goal: how much the participants intended to exercise in the following week. After completing the same handwriting task (using only the phrases 'I will' or 'will I?') the participants were asked to fill out a psychological questionnaire to measure intrinsic motivation. Motivation is the force that drives us to achieve our goals. Motivation is either extrinsic (it depends upon rewards such as money, prestige or power) or intrinsic (it comes from desire to perform the task for the enjoyment it provides). Senay got the same results in this real-world scenario: those who wrote 'will I?' expressed a greater commitment to exercise regularly than those who wrote 'I will'.

By asking ourselves a question we are more likely to build our own intrinsic motivation. It seems it's a more promising way of achieving our goals. Senay's research also shows how language provides a window between thought and action. It also shows the way we talk about our behaviour can predict future action.

The forceful 'I will' is not the way to go about achieving our goals. 'Will I?' makes the activity intrinsically attractive by placing it in the future. We are more likely to do it. When it comes to achieving goals, arguably Arianna Huffington, an American author and syndicated columnist, has the right advice: in her book *Thrive* she says that the best way to achieve a goal is to drop it.

Will you drop your goals? Can't decide? If you have your dog or cat around—or even just in mind—may help you achieve this goal (of answering this question) as well as generate more goals and feel more confident about achieving them. In an experiment, researchers assigned participants in one of the three conditions: a pet nearby, simply thought about a pet and no pet presence. Those who had their pet in their room or in their mind identified more goals and showed confidence in achieving them. In a second experiment the participants performed a distressing mental task while their blood pressure was measured. Participants with their pet in the room or in their mind had lower blood pressure than the participants with no pet presence. Can't keep up with your new year resolutions? Think of your dog or cat.

Siddhartha continues:

> 'You, oh venerable one, are perhaps indeed a searcher, because, striving for your goal, there are many things you don't see, which are directly in front of your eyes.'
>
> Hermann Hesse, *Siddhartha*

A tweet from science and philosophy worth retweeting: the future is not a closed answer; it's an open question.

Chapter 5

The pain of unpopularity

learn resilience is the best strategy to cope with social rejection/
unlearn the pain of rejection cannot be reduced

> The avant-garde in every field consists of the lonely, the friendless, the uninvited. All progress is the product of the unpopular.
>
> People in love—with nurturing, attentive non-movie-star parents—they would never invent gravity. Nothing except deep misery leads to real success.
>
> – Madison Spencer in her email 'Now, Voyager!' from the Great Beyond
> Chuck Palahniuk, *Doomed*

While looking at a world map have you ever wondered whether the shapes of Africa and South America look as if they could fit together like the pieces of a jigsaw puzzle? Browsing through at an atlas that a friend had received as a Christmas present in 1910, Alfred Wegener, a 30-year-old German meteorologist, was struck by this fact. This simple observation led him to develop his theory of continental drift: the continents had once been joined together in a giant supercontinent that began to break away about 200 million years ago into the continents we know today, which slowly started drifting into their current positions. When he presented the theory in 1915 in his book *The Origin of Oceans and Continents*, it became the most controversial, derided and ridiculed book in the history of geology. Some prominent scientists labelled the book 'utter damned rot!', and questioned Wegener's 'scientific sanity'.

History of science is full of incidences when new ideas had been ridiculed not only by fellow scientists but also by the public. In the early 1530s, uneducated and pig-headed people merely poked fun at Copernicus' idea that the Earth moves around the sun; but learned people, especially religious leaders, were full of venom and denounced him as 'an upstart astrologer, the fool who wanted to turn the whole science of astronomy upside down'. Two centuries later, William Harvey's findings on the circulation of blood were rejected by his fellow physicians because they maintained that his ideas had not cured a single patient, and the public ridiculed him as a 'crack-brained circulator' ('circulator' is the Latin slang for 'quack'). Rachel Carson's prophetic book, *Silent Spring*, triggered the modern environmental movement; but when it was published in 1962 a *Time* magazine review panned the book for 'oversimplifications and downright errors' and multinational chemical companies attacked her as a 'hysterical woman' unqualified to write such a book.

Even Aristotle didn't escape unpopularity. A year before his death in 322 BC he was charged with impiety, the lack of reverence for religion. Rather than suffer the fate of Socrates he fled Athens saying that he would not allow Athens to sin twice against philosophy. Aristotle believed that reasoning and investigation are worth doing because of what they might deliver—power, wealth or popularity. But they should not be avoided for ridicule and rejection they might bring. These words might bring some solace to scientists but not to ordinary people like you and me when faced with criticism and rejection in our professional or social lives.

Responses to rejection vary considerably but many people do abandon their careers or change lifestyles because of their inability to cope with rejection. The best strategy, of course, is resilience—a quality that can be learned. Resilience has become the focus of cognitive psychology, neuroscience and business research in recent years. The research has identified skills you can develop to boost your resilience. Your progress—and eventual success—in any endeavour you undertake may hinge on your resilience.

Stand up to stress. The hippocampus, the region of the brain that is closely linked with the limbic system that controls emotions and motivation, is most susceptible to stress hormones such as cortisol and noradrenaline (also called norepinephrine). But dopamine makes it happy. Even a single severe stress episode can destroy newly created neurons in the hippocampus. Chronic stress can kill off old neurons in the hippocampus. Research on post-traumatic stress after 9/11 terrorist attacks shows that hippocampus of some survivors had shrunk to the size of those elderly people with dementia.

The simplest ways to avoid stress is to include physical exercise in your daily routine. 'Changing routines can also interrupt a dark period,' says Elaine Fox, author of *Rainy Brain, Sunny Brain*. 'Take a different route to the grocery store, for example, or call up a friend out of the blue.'

Remain hopeful. When you have a setback, don't lose sight of your goal. Don't blame yourself for your misfortunes. Look for the silver lining in the cloud that has darkened your life.

Optimism is like the placebo pills. Physicians have long known that sugar pills disguised as medicines— placebos—can help some patients. Thalamus, a region of the brain that acts as the gatekeeper by relaying sensory information, releases pain-reducing chemicals such as dopamine after a placebo is given. Believing in a cure not only makes you feel better, it can lead to dramatic bodily changes. Belief is one of the most powerful medicines.

Overall, optimism is good for you, but there are limits to seeing the glass half full. Unrealistic optimism, however, could lead you to ignore real dangers—not seeing that glass as really empty.

Make social connections. Strengthen your relationship with family, friends and work colleagues. Studies show that social networking sites like Facebook are likely to have positive effects when used by people with low self-esteem or depression.

However, studies also showed that those with low self-esteem also check their Facebook pages more frequently than normal. Steven Southwick and Dennis Charney, authors of *Resilience: The Science of Mastering Life's Greatest Challenges*, offer the following valuable advice: 'In your social circles, look for people who recover quickly from hardship whom you could learn from. Members of your own family, colleagues, teachers, coaches, or even historical figures or fictional characters could also serve as resilient role models.'

Consider and take decisive actions. 'In the day of prosperity be joyful, but in the day of adversity consider,' says the Bible (Ecclesiastes 7). It's not often that the ancient wisdom coincides with new thinking. Consider: not to ignore your problems, keep things in perspective, take decisive actions, change your goals, make every day meaningful.

Feeling sleepy while reading this book? Pssst! If the great philosopher hears it he would insist that you use his contraption while reading a book. Aristotle always placed beside his bed a brass dish, and when he lay down to read and rest, he would hold in his hand extended over the dish a brass ball. When he was overcome by sleep, the ball would fall into the dish, and the noise would immediately awaken him.

The right contraption to study the effects of rejection on the brain is the fMRI (functional magnetic resonance imaging). The brain has a pain-control system that damps down the perception of unwanted stimuli. The brain region called the dorsal anterior cingulate cortex, which resides about 2.5 centimetres behind your forehead, is the key regulator of pain signals: it lights up when we feel pain, and also when we are emotional. Our bodies respond to rejection like they do to physical pain: those who are rejected show increased activity in this brain region; soothing the body's response to rejection eases the pain and slows down activity. Spurned lovers know now why it feels like hell when they are dumped.

In an experiment C. Nathan DeWall, a social psychologist at the University of Kentucky, recruited 62 healthy volunteers who took two tablets of either paracetamol (also called acetaminophen; sold as Panadol, Tylenol and many other brand names) or a placebo every day for three weeks. Each evening the participants completed a questionnaire measuring their feelings of rejection. DeWall and colleagues found that the participants who were taking paracetamol tended to report less hurt feelings and more resilience over time than the participants who were taking placebos. Subsequent fMRI scans revealed that those who took paracetamol showed less activity in the dorsal anterior cingulate cortex compared to those who took the placebo.

The pain of social isolation has an evolutionary advantage. This advantage can be traced back to early in our history as a species when we survived and prospered only by living in groups to provide mutual protection and help. If you were socially isolated, you were likely to die. It's important to be able to feel the pain of social isolation: it's a signal to renew social relationships we need to survive and prosper. After millions of years this pain of rejection still hurts. Advises DeWall: 'When you are rejected or excluded, the best way to deal with it is to seek out other sources friendship or acceptance.' In other words, build up your resilience.

Why bother about taking paracetamol to reduce the pain of rejection? Instead, laugh—laugh at yourself or anything that comes to your mind. Laughter gives our brains a rush of endorphins, the feel-good hormones. 'The burden of self is lightened/when I laugh at myself,' writes Rabindranath Tagore in *Fireflies*, a collection of epigrams and short verses.

Chapter 6

Turn up the wattage of your smile

learn to smile more, 'a smile that just seems so genuinely sweet'/ *unlearn* to scowl, 'You don't scowl. Improves your looks a lot.' (Suzanne Collins, *The Hunger Games*)

> 'You girls,' said Miss Brodie, 'must learn to cultivate an expression of composure. It is one of the best assets of a woman, an expression of composure, come foul, come fair. Regard the Mona Lisa over yonder!'
> All heads turned to look at the reproduction, which Miss Brodie had brought back from her travels and pinned on the wall. Mona Lisa in her prime smiled in steady composure even though she had just come from the dentist and her lower jaw was swollen.
>
> Muriel Spark, *The Prime of Miss Jean Brodie*

When did you first smile? When you were 4 to 6 weeks old (probably during a feeding). When you were 40 to 48 weeks old you were responding to smiles of others. The duration of a smile is very short: typically lasting from two-thirds of a second to four seconds. Those who witness this transient spark of emotion often respond by smiling back. It takes more muscles to frown than to smile. The 'smile muscles' work very hard giving our brains a rush of endorphins, the hormones that are responsible for our general sense of wellbeing. Smiling is like laughing, singing, listening to good music, meditating or even eating chocolate, which also help release feel-good endorphins in the brain.

Can you spot a genuine smile? Of course, you can. You can simulate a smile but only certain involuntary muscles produce a spontaneous expression of positive emotion. A genuine smile is the 'Duchenne smile', for psychologists at least. In the 1860s a French neurologist, Guillaume Duchenne, studied facial expressions by triggering muscular contractions with electrical probes, which he recorded with a camera. He showed that only the 'sweet emotions of the soul' forces both involuntary and involuntary smile muscles to contract to produce a genuine smile. The voluntary muscle raises the corners of the mouth and the involuntary muscle raises the cheeks and produces crow's feet around the eye. In a fake smile, or say-cheese smile, only the voluntary muscle contracts since involuntary muscles are not in our conscious control.

Several studies have shown engaging effects of a genuine smile. In a 30-year study, researchers matched college yearbook photos of women who displayed the genuine smile with personality data collected when these women were at age 27,

43 and 52. They found that these women had higher levels of wellbeing and marital satisfaction three decades later when they were in their early 50s.

A recent study that builds up on this finding involved examining a variety of participants' photos from childhood through early adulthood. The study also examined participants' college photos. In both cases, divorce was predicted by the brightness of smiles of participants in their photos. One in four people whose smiles were weakest in their college photos ended up divorcing, compared with one in 20 of those with brightest smiles. The same pattern was true even for those pictured at an average age of ten.

Another study rated smiles of professional baseball players in an old yearbook. They then rated each player's age at death (some players were still alive at the time of the study). Their results? Players who sported genuine smiles were only half as likely to die, in any given year, as those who had not.

It's fascinating to learn that a photograph not only captures passing emotions of the moment but also the future. Moral of above studies: smile naturally and brightly when a camera lens is staring at you.

Looking for excuses to smile more?

A smiling face makes you more memorable. In a Duke University study volunteers were 'introduced' to a number of strangers by showing them strangers' photos and telling them their names. Using brain scans, the researchers found that both learning and recalling the names of smiling faces lighted up sensory reward centres in the brain. 'We are sensitive to positive social signals,' explains the lead researcher Roberto Cabeza. 'We want to remember people who are kind to us. In case we interact with them in the future.'

A smiling face makes you more attractive. Research—and anecdotal evidence—supports the claim that smiling enhances our attractiveness. The enigmatic Jay Gatsby in F. Scott Fitzgerald's *The Great Gatsby* had 'one of those rare smiles with a quality of eternal reassurance in it ... and assured you that it had precisely the impression of you that, at your best, you hoped to convey.'

Smiling makes you happy. As early as 1872 Charles Darwin observed in *The Expression of the Emotions in Man and Animals* that emotional responses influence our feelings. 'The free expression by outward signs of an emotion intensifies it,' he wrote. 'On the other hand, the repression, as far as this is possible, of all outward signs softens our emotions.' Psychologists do not yet know why our facial expressions influence our feelings, but we do know our faces open a window into our minds not only to others but also to ourselves.

Cosmetic botox injections compromise ability to frown. Psychologists at the Cardiff University in Wales have found botox recipients feel happier and less anxious in general. 'It would appear that the way we feel emotions isn't just restricted to our brain—there are parts of our bodies that help reinforce the feeling we're having,' says Michael Lewis, a co-researcher. This idea also works the opposite way. Studies have shown that people who frown during an unpleasant medical procedure report feeling more pain than those who do not.

Cognitive scientists Tara Kraft and Sarah Pressman of the University of Kansas recruited 169 volunteers for their two-phase study. During the training phase, participants were instructed to hold chopsticks in their mouths in such a way that they engaged facial muscles to create a fake smile, a genuine smile or no smile at all. In the test phases, participants worked on stress-inducing activities such submerging a hand in ice water while holding chopsticks in their mouth as they were taught in training. The researchers measured participants' heart rates and self-reported stress levels during the testing phase.

Their findings show that smiling during brief periods of stress can help reduce the intensity of body's response to stress. 'The next time you are stuck in traffic or are experiencing some other type of stress,' says Pressman, 'you might try to hold your face in a smile for a moment. Not only will it help you "grin and bear it" psychologically, but it might actually help your heart health as well!' Put a pencil or chopstick between your lips the right way, and you're likely to feel happier. But not at the red traffic light.

Smile to make someone else's day. A Scandinavian study confirms the ages-old adage it is actually hard to frown when someone else is smiling. We have a tendency to mimic emotions of those around us.

Smile genuinely and widely as you read this verse from 'Little Things', a poem by an unknown 19th-century poet:

> It was only a sunny smile,
> And little it cost in the giving;
> But it scattered the night
> like the morning light,
> And made the day worth living.

Chapter 7

Hip hip hippocampus

learn how exercise jogs the brain and makes you smarter/
unlearn that physical exercise is only about losing kilos and living skinnily

> She liked that show. She found it soothing to get caught up in a brightly coloured, plastic world where all that mattered was how much you ate and exercised, where pain and anguish were suffered over no greater tragedy than push-ups, where people spoke intensely about calories and sobbed joyfully over lost kilos. And then they all lived happily, skinnily ever after.
>
> – Rachel watching *The Biggest Loser* on TV
> Liane Moriarity, *The Husband's Secret*

'Why are you talking about some dumb TV show and not me?' Pardon me; a cheeky hippocampus is nudging me to talk about things that would make it—and you—shout hip hip hooray. No, it won't sit down on the couch with you. Couch potato isn't; the lack of exercise makes it shrivel.

First, it wants to tell a little anecdote: When a traveller asked Wordsworth's servant to show him her master's study, she answered, 'Here's his library, but his study is out of doors.'

Walking in the fields and woods made Wordsworth think clearly, but he didn't know that regular brisk walking (or any other aerobic exercise such as jogging, swimming, cycling, dancing) lowers risks of dying from heart disease, stroke, diabetes, some forms of cancer and other ills. You do.

Where's that exercise bike? Aerobic exercise also makes you smarter by boosting your mood, memory and learning. That's where the hippocampus comes in. This seahorse-shaped centre of learning and memory in the brain is also closely linked with the limbic system that controls emotions and motivation. So, it truly takes care of your wellbeing. Your wellbeing is inextricably linked to the wellbeing of your hippocampus.

When you walk or exercise, neurons in the hippocampus rev up which in turn improve your cognitive abilities. The revved up neurons also lift up the mood by releasing neurotransmitters like serotonin, noradrenaline and dopamine—the same neurotransmitters that antidepressants and drugs for attention-deficit hyperactivity disorder act on. Why bother swallowing Prozac or Ritalin? 'So, if you are having a mental block, go for a jog or hike,' advises Justin Rhodes, a psychologist at the University of Illinois. 'The exercise might help you pull out of your funk.'

It might also help increase the size of your hippocampus, which tends to decrease in old age. It used be thought that ageing was a one-way process that was going the wrong way, but we know now that's not the case. Studies show that hippocampus is larger in elderly people who walk, jog or engage in aerobic exercises regularly. Any shrinkage in hippocampus can lead to Alzheimer's disease and dementia in general. Obviously, motivation and drive suffer in most types of dementia. Aerobic exercise can improve the brain's decline or at least it can slow it down.

For six months researchers from Canada and the Netherlands studied 86 women aged between 70 and 80 years with mild memory problems. The study, published in the *British Journal of Sports Medicine*, focused only on women to avoid potential gender differences in how the brain responds to exercise. The women who were living independently at home were assigned to twice weekly hour-long sessions of aerobic training such as brisk walking; resistance training such as lunges, squats and weights; or balance and muscle toning exercises (this one was the control group). The size of their hippocampus was measured by fMRI scans at the start and the end of the study.

The results showed that the hippocampus was significantly bigger in those who completed six months of aerobic exercises compared with balance and muscle and toning exercises. Resistance training group did not benefit much. The relationship between the size of the hippocampus and mental performance is complex but at the very least aerobic exercise seems to slow down the shrinkage of hippocampus in older people who are at the risk of developing dementia.

Exercise benefits the brain in more than one way. Exercise pumps more blood to the brain, which gives more oxygen to the neurons and thus making them better nourished. Exercise also causes the release of certain proteins known as growth factors. As the levels of these proteins build up neurons start to branch out and start building new connections in the hippocampus. These new connections signify a new fact or skill, which has been learned and stored for future use. As we age, individual neurons start to die. This loss is not permanent: the brain can make new neurons. Again, these proteins play a role in growing new neurons—and exercise helps in building up its levels by increasing blood volumes.

Mens sana in corpore sano (A healthy mind in a healthy body). That's what your grandmother said; that's what science says. To keep revving up those neurons in your brain all you have to do is 150 minutes of moderate, or 75 minutes of vigorous, exercise every week. For older people, brisk walking is the best exercise.

Regular physical activity not only revs up neurons in your brain and improves cognitive abilities that shape your IQ it also benefits the body in numerous other ways. It expands lung capacity; protects the body from inflammation; reduces the risk of developing or dying from heart disease and stroke; improves the body's response to insulin and decreases the risk of type 2 diabetes; reduces the risk of breast, colorectal and other cancers; reduces the risk of falls and fractures and increases muscular efficiency; can turn on or off specific genes; and, of course, keeps the excess weight in check. What else you want to live happily and skinnily?

You are never too old—or too young—to reap the benefits of exercise. 'We need to have kids moving every day, not just because it makes sense health-wise, but because it raises test scores,' exhorts John Ratey, a psychiatrist at Harvard Medical School. Little Winnie-the-Pooh, the bear with no brain, knew that exercise would make him smarter as well ('A bear, however hard he tries, grows tubby without exercise.' A. A. Milne, *Winnie-the-Pooh*).

Like Wordsworth, make outdoors your study and stroll in the footsteps of Mrs Wells 'who was not insensible of the myriad benefits of daily perambulation, and who often said that there was nothing she liked better than a stroll' (Eleanor Catton, *The Luminaries*). After your perambulation you may shout hip hip hippocampus.

Chapter 8

Older, a bit slower but definitely smarter

learn an ageing brain is still pretty smart/*unlearn* the popular belief that declining brainpower is part of growing older

> *Dementia.* Ruth puzzled over the diagnosis: How could such a beautiful-sounding word apply to such a destructive disease? It was a name befitting a goddess: Dementia, who caused her sister Demeter to forget to turn winter into spring. Ruth now imagined icy plaques forming on her mother's brain, drawing out moisture.
>
> Amy Tan, *The Bonesetter's Daughter*

When older people begin to mutter whatshisname or whatshername, we all assume that's what happens in old age: our brainpower declines as we grow older. Onomastic aphasia, the medical name for whatsisname condition (the name is on the tip of your tongue but can't recall), conjures up painful images of dementia: memories of the past slowly slipping away in old age.

Without a doubt our brains change as we age from early 20s to 65 and beyond. The physical changes include loss of neurons and nerve fibres and significant shrinking of some brain regions such as the hippocampus. Along with these changes comes decline in our ability to think and reason. It's still a matter of debate when cognitive decline begins. Some studies suggest after 60; others as early as 45.

Michael Ramscar, a linguistics researcher at Universität Tübingen in Germany, was 45 when he read in a research paper that our vocabulary declines after age 45. That didn't made sense to him. 'Ninety-nine per cent of the people I look up to intellectually,' he wondered, 'are older than I am.' A few years later, a major study led by Ramscar has revealed that the loss of neurons does not play a significant role as we age; and fading brainpower is not an inevitable part of growing older. The study has turned the popular belief on its head.

Ramscar ascribes the popular belief, in part, to Greek mythology. Eos, goddess of dawn, begged Zeus to grant immortality to Tithonus, a mortal whom she had married. Zeus agreed to this request. But she forgot to ask also for perpetual youth dooming Tithonus to an eternity of physical and mental decay. Ramscar remarks that Tithonus' account of ageing echoes loudly in brain-science literature, which portrays old age as a protracted episode in mental decline, in which memories dim, thoughts slow down and problem-solving abilities diminish.

He suggests that many of the assumptions scientists currently make about 'cognitive decline' are seriously flawed and, for the most part, formally invalid. He agrees that our brains work slower in old age but only because we have stored more information over time. 'The brains of older people do not get weak,' he says. 'On the contrary, they simply know more.' Older brains are so jam-packed with knowledge that they simply take longer to retrieve the correct bits of information. This brimming store of knowledge helps older brains to compensate any loss related to ageing.

Ramscar and colleagues used standard computer models of human cognition processing but they added new information to model's memory bank to simulate the accumulating experience as we age. In one of their experiments they used paired-associate-learning, a test often used to measure our ability to learn and recall new information. In this test, people learn to connect words in pairs between word cues (for example, *baby, jury*) and word responses (*cries*, *eagle*). People perform worse on this task as they grow older, supporting the conclusion that learning ability declines with age. The researchers found that younger adults did better on 'easy' pairs such as *baby-cries* than 'hard' pairs such as *jury-eagle*, but older adults also understood which words don't usually go together. When the researchers examined performance on this test across a range of word pairs that go together more or less in English, they found older adult's scores to be far more closely attuned to actual information in hundreds of millions of words of English than their younger counterparts.

Ramscar and one of colleagues on the research team, Harald Baayen, write in *New Scientist* that performance on tests like paired-associate-learning are not evidence of cognitive or physiological decline; they are evidence of continued learning and increased abilities. 'People who believe their abilities can improve with work have been shown to learn far better than those who believe abilities are fixed,' they write. 'It is sobering to think of the damage that the pervasive myth of cognitive decline must be inflicting.'

The Ramscar study has also come up with an explanation for whatsisname condition. There is a greater variety of given or first names than there were two generations ago. This means the number of different names we learn over our lifetimes has increased dramatically. Locating a name in memory, therefore, is far harder than it used to be. It's true even for computers; that's why we need supercomputers.

Don't worry if you can't recall the name of the author of this book. As long as you believe in yourself, you're a superbrain. Anyway, a trick from neuroscientists to remember a name: repeat the name to yourself every 10 minutes. This idea is based on experiments showing that synapses become permanently strengthened if a stimulus is applied at about 10-minute intervals. It's all about moving information from the brain's short-term storage to long-term storage.

A study by Rosalyn Moran, an American researcher in the neurobiology of ageing, and her colleagues also add to the new thinking that age-related decline in brainpower is just a myth. They gave the mismatch negativity (MMN) test to 97 healthy volunteers aged 20 to 83. In a typical MMN test, a sound is played again and again, and then an 'oddball' sound is played instead. This triggers a change in brain waves, which can be measured using electroencephalography (EEG). In older people the MMN response is generally slower and weaker suggesting cognitive decline. The Moran team's results, however, showed that while the older brain learns to expect the standard MMN sound, it does so quickly and less intensely, and its surprise at 'oddball' sound is also muted.

The results of the Ramscar and Moran studies show that so-called cognitive ageing reflects accumulating experience rather than deterioration. Erik Erikson, the pioneering development psychologist who researched life phases, also said decades ago that ageing is a process of development and progress, not decline.

Well, old dogs need not to learn new tricks. But they must heed the good ol' advice to deter dementia: don't smoke, try to stay in shape by exercising regularly, and keep an eye on your blood pressure and cholesterol levels. What's good for the heart is good for the brain. A diet rich in fish and vegetable oils, less added sugar, non-starchy vegetables and fruits with low glycemic index is the best. This Mediterranean diet doesn't forbid small pleasures such as a glass of red wine or beer (have a bite to eat first), and a small bar of dark chocolate. Old dogs know what moderation means. Also, banish the thought of 'retirement'. Engage in meaningful work whether that means working for pay, volunteering, community activities or simply looking after grandchildren.

Ernest Hemingway writes in *A Farewell to Arms*: 'No, that is the great fallacy: the wisdom of old men. They do not grow wise. They grow careful.' No, 'Papa' Hemingway, they grow careful and wiser.

Chapter 9

One and only one brain

learn left and right hemispheres of the brain work together in complex and collaborative ways/*unlearn* differences in left brain/right brain dominance can help explain individual differences among learners

> 'Isn't that part of the brain associated with numbers and mathematical capacity?' Jonasson said.
> Ellis shrugged. 'Mumbo jumbo. I have no idea what these particular grey cells are for.'
>
> Stieg Larsson, *The Girl Who Kicked Hornets' Nest*

You have probably heard that rational, analytical and logical thinkers such as engineers and mathematicians are left-brained; and creative and emotional people such as artists and musicians are right-brained. This widely held view is based on the assumption that rationality, logic and verbal skills are located in the left hemisphere of the brain, while creativity, emotions and visuospatial skills are located the right hemisphere. This thinking is not simply true.

The erroneous thinking that information is processed in different ways in the two hemispheres of the brain is still reflected in our schools: the best teaching techniques for left-brained people should involve verbal instructions, talking and writing, and multiple-choice questions; while demonstrated instructions, drawing and manipulating objects, and open-ended questions are best for right-brained people. This notion has led to the idea that education programs should synchronise the two hemispheres by including both left-brained and right-brained activities. 'Show and tell' activities of your early schools days are the result of this thinking: instead of only reading a 'left-brained' text, your teacher also showed pictures and graphics to stimulate your right hemisphere.

The left brain/right brain myth can be traced back to the days of 19th-century craze of phrenology. Phrenologists studied shape and size of the head to determine a person's character and mental abilities. They believed that different mental functions were located in different organs of the brain, and the growth of the various organs was related to the development of associated mental abilities. As this growth would be reflected in the shape of the skull, personality traits could be determined by reading bumps and depressions on the skull. For example, if you move your finger on the back of your neck, you will notice a bump formed by the base of your skull. This bump, according to phrenologists, defined the attachment of sexes to each other. In 1844 this mumbo jumbo became popular when a book described the two hemispheres of brain as independent parts having an independent way of thinking. The idea even found way into Robert Louis Stevenson's famous story *The strange case of Dr Jekyll and Mr Hyde* published in 1886.

In the 21st century we have sophisticated brain-imaging techniques that reveal less romantic sides of the brain: there is no evidence that the left brain is 'mathematical' and the right brain 'musical'. Yes, the brain is divided into two hemispheres. They look almost identical anatomically, but they are not independent. They are connected by thick bundles of nerve cells that carry information from one side to the other. The two hemispheres differ not so much in what they do, but in how they process tasks. The left hemisphere is better at details (such as recognising a particular face in a crowd), whereas the right hemisphere is better at dealing with a general sense of space (the relative positions of people in a crowd). In the case of language, for example, the left hemisphere focuses at step-by-step processes, such as grammar and word generation, whereas the right hemisphere focuses at feeling a rhythm, such as intonation and emphasis of speech. Similarly, brain-imaging studies on emotion provide no scientific support for the idea that the brain's emotional functions are confined to the right hemisphere.

There are no specific 'left brain' or 'right brain' cognitive functions. Both hemispheres work in concert with each other on every cognitive task, whether we are reading, painting or solving an algebra equation. The brain is remarkably adaptive and children who have had hemispherectomy—in which half of the brain is removed—the remaining half overtakes most of the functions of the missing half. Yes, half a brain is better than a whole brain divided into mythical left brain, right brain. Whether you are young or old, learning actually strengthens and creates new connections between neurons.

A 2012 study by educational neuroscientist Sanne Dekker of VU University Amsterdam and her colleagues shows that 91 per cent people in the UK and 86 per cent in the Netherlands believe in the neuromyth that differences in hemispheric dominance (left brain, right brain) can help explain individual differences amongst learners. Surprisingly, the researchers' sample comprised 242 primary and secondary school teachers who were interested in the neuroscience of learning. The researchers say that 'teachers who are enthusiastic about the possible application of neuroscience findings in the classroom, often find it difficult to distinguish between pseudoscience from scientific facts'.

Parents and teachers need to be careful to base learning and teaching on 'just discovered' science. A report on a research study you may read in the popular media doesn't mean the study result has been automatically stamped 'proven by science'. Science advances unpredictably, not linearly; one scientific study often disputes the other, sometimes followed by the third that disputes both. An idea is labelled truly 'scientific' only when it has earned the consensus among the majority of scientists in that particular field.

It's time we used our whole brains to learn that like Chinese Yin Yang symbols the two hemispheres of our brains are in perfect harmony. There is no program or technique that can boost capabilities of your left or right brain. Similarly, no scientific study supports the claims made by 'whole brain' training programs. Why waste money on brain-training programs to exercise your brain, when you can exercise your brain on your own for free by learning a new language or learning to play bridge, chess or a musical instrument?

Chapter 10

The brain goes to school

learn to your surprise that many teachers peddle neuromyths/
unlearn all teaching practices in your children's school are based on sound educational neuroscience research

> 'I think so,' said Professor McGonagall dryly, 'we teachers are rather good at magic, you know.'
>
> J. K. Rowling, *Harry Potter and the Deathly Hallows*

'And very good at perpetuating myths.' No, Professor McGonagall didn't say that but is there a difference between magic and myths when it comes to teaching children?

Ready for a simple test? Take it yourself or give it to teachers of your children. True or false?

1. We only use 10 per cent of our brain.
2. When we sleep, the brain shuts down.
3. Brain development has finished by the time children reach secondary school.
4. The brains of boys and girls develop at the same rate.

5. Mental capacity is hereditary and cannot be changed by the environment and or experience.
6. If children do not drink sufficient amount of water (that is, six to eight glasses a day) their brains shrink.*
7. Children are less attentive after consuming sugary drinks and/or snacks.*
8. It has been scientifically proven that fatty acid supplements (omega-3 and omega-6) have positive effect on academic achievement.*
9. Children must acquire their native language before a second language is learned. If they do not do so neither language is fully acquired.*
10. There are critical periods in childhood after which certain things can no longer be learned*
11. Girls are better at reading, but boys dominate maths and science.
12. People are either 'right brained' or 'left-brained' and these differences can help explain individual differences amongst learners*
13. Short bouts of exercise can improve integration of function of left and right hemispheres of the brain
14. Individuals learn better when they receive information in their preferred learning style (visual, auditory, read-write, kinaesthetic)*
15. Environments that are rich in stimulus improve the brains of pre-school children*

16. Exercises that rehearse coordination of motor-perception skills can improve literacy skills*
17. Listening to classical music makes you smarter
18. Learning problems associated with developmental differences in brain function cannot be remediated by education.*
19. Children who receive training to boost emotional intelligence, learn more effectively and mature more quickly
20. Regular aerobic exercise can improve mental function

Add up your score. Statements 1 to 19 are false; 20 is true.

A survey of hundreds of primary and secondary school teachers in the UK and the Netherlands by educational neuroscientist Sanne Dekker of VU University Amsterdam and her colleagues revealed the shocking truth that most teachers fail to distinguish pseudoscience from scientific facts. The teachers surveyed were not ordinary teachers; they were enthusiastic about possible application of neuroscience findings in the classroom. The teachers were asked to categorise 32 statements about the brain as true and false (some of the statements are included in the list above but they have been rephrased). Embedded among these were 15 neuromyths about education (including the 10 items marked with an asterisk in the list above). Results showed that, on average, teachers believed 49 per cent of the neuromyths. Around 70 per cent of general knowledge statements about the brain were answered correctly. Teachers who read popular magazines achieved higher scores on general knowledge questions; and amazingly these teachers showed an increased belief in neuromyths about education.

'Possessing greater general knowledge about the brain does not appear to protect teachers from believing in neuromyths,' Sanne Dekker and colleagues write. They also stress that it is important to examine where teachers' incorrect ideas originate. From books, colleagues, or commercial companies spruiking brain training programs? They also urge neuroscientists to check the way their research is portrayed in the media. Newspaper headline writers and blogging enthusiasts please take note of these findings.

Not every research finding automatically becomes truly 'scientific'; for that label it has to earn the consensus among the majority of scientists in that particular field. The results of brain-imaging studies may appear real and reliable, but these studies are usually done on very small samples because the machines are complex and expensive. Scientists are aware of the limitations of their studies, but in the hands of popular media the results become 'scientific proofs'.

As cognitive neuroscientists make advances in our understanding of how the brain functions, the challenge for teachers and parents is to integrate this new knowledge of the mind into schooling. But the task is not simply reading about a new research finding in the media and applying it into the classroom. Students are not guinea pigs. Parents and teachers must be careful about the claims supposedly based on neuroscience.

A review by researchers at Newcastle University in the UK has identified 71 learning styles most of which are not backed by credible evidence. The most popular is VARK (visual, auditory, read-write, kinaesthetic), which argues that every child has an individual learning style and they learn, receive, process and retain information when it is delivered in a mode suited to their style. It advocates pictures for films and signs for 'V' children, traditional written material for 'R' children, and so on. There is no scientific evidence to support VARK.

Another concept that needs closer examination is that of 'emotional intelligence' made popular by American psychologist Daniel Goleman in his 1995 book, *Emotional Intelligence*. He called qualities such as self-control, persistence and motivation emotional intelligence or EQ. The idea was immediately picked up by management gurus who used it to assess and nurture employees' behaviour and management skills. Educators also jumped on the bandwagon believing that if students received training to boost their EQ it will improve their self-awareness and self-motivation and they will learn more effectively. But there is no neurological evidence to support the relationship between emotional intelligence and general intelligence.

In 2011, Goleman wrote in *Time* magazine: 'Have you heard? They say that your EQ counts more than your IQ for success. In fact, they say, EQ accounts for 80% of success. As the person who wrote *Emotional Intelligence*, the book that put the concept on the map, I can tell that they are dead wrong. This and other myths about emotional intelligence constantly float around the blogosphere and get spouted by management consultants.' Why do people see EQ as a magical alternative to IQ? Goleman believes that for some, it may be a consolation for poor performance in schools; for others a code for humanising the workplace. EQ also offers a politically correct argument for more women in leadership. There is no hard data yet to support these reasons. 'The slow march of research lags far behind the hype of EQ marketers,' Goleman says.

On the plus side is the idea that teenagers do not like early starts because they have a natural sleep pattern that leads to late-to-bed, late-to-rise cycle. Some American schools have adopted this credible research by starting their classes later in the morning. Systematic reviews of the later start times have found it beneficial.

The new discipline of educational neuroscience or neuroeducation hopes to bridge the gap between neuroscience research and education practice by connecting cognitive neuroscientists who study learning and teachers who want to make sense of the new research. But neuroeducation researchers also face hurdles as they have no tradition of applying neuroscience findings to the practice of teaching. 'It is difficult to provide benchmarks for good work,' worries Professor Howard Gardner of Harvard Graduate School of Education. 'We must begin to establish that tradition.'

Until neuroeducation brings some scientific rigour to teaching, teachers and parents are on their own. At least, they should know that neuroscience places no constraints on any child's ability to learn. It's up to them to help children realise their 100 per cent potential by offering a learning environment that bears the true stamp of 'approved by neuroscience'. Oh, yes, no neuroscientist has ever found that 90 per cent of the brain is perpetually on vacation and yet the myth of 10 per cent brain perpetuates.

Chapter 11

Silencing sirens of homework police

learn helping your children with homework doesn't make them smarter/*unlearn* to be homework police

> I got good grades with minimum effort. My accountant father did my math homework and prepared test crib sheets for me. I was free to read and dream away my off-school hours.
>
> – novelist James Ellroy talking about his school days
> James Ellroy, *My Dark Places*

Homework has always been a controversial issue for parents and teachers. In the 1980s when I was editor of an education magazine I remember publishing numerous stories on homework, each of which invariably generated a stream of letters from concerned readers. The topics included: parents demand homework; children have way too much homework, it's too little or it's not intellectually demanding; homework gives parents one-to-one tutoring opportunities which benefits children; homework encourages the ability to work independently in a self-motivated way; and so on. Look at any print or online magazine today—not only education and parent magazines but also prestigious magazines like *Time* and *The Atlantic*—and you will find the same old issues being discussed ad nauseam.

The conventional wisdom—children do better when parents are actively involved in their children's education, meeting with teachers, helping them with their homework, and scores of other things—has now been challenged by a groundbreaking study by American sociologists Keith Robinson and Angel L. Harris. Their longitudinal survey of American families across socioeconomic and ethnic groups spanned from the 1980s to the 2000s and assessed more than 60 measures of participation, at home and in school.

In *The Broken Compass: Parental Involvement with Children's Education*, they conclude that parental involvement is overrated. 'In fact, most form of parental involvement, like observing a child's class, contacting a school about child's behaviour, helping a child decide a child's high school course, or helping a child with homework, do not improve student achievement,' they say. 'In some cases, they actually hinder it.'

Robinson and Harris were quite startled by what they found when they examined whether regular help with homework had a positive effect on children's academic performance. 'Regardless of a family's class, racial or ethnic background, or a child's grade level, consistent homework help almost never improved test scores or grades,' they say. 'Most parents appear to be ineffective at helping their children with homework. Even more surprising to us was that when parents regularly helped with homework, kids usually performed worse.'

How could parental help with homework bring the scores down? Probably many parents have forgotten, or never truly understood, the things children learn in school. Nevertheless, the study revealed that Asian children, including Chinese, Korean and Indian, benefitted from regular help with homework, but this benefit was limited to the grades they got during adolescence, it did not affect their test scores.

What should parents do? Their answer is simple: 'They should set the stage and then leave it.' But what kind of stage? If children whose brains are still developing are constantly juggling between homework and texting or listening to iTunes and so forth there will never be enough depth and times spent on homework to go as deep or as far as they might have. Yet, parents need not be homework police (hovering like helicopter over their children while they are doing their homework) or enthusiastic homework helper (rushing to help children when they are unable to solve a problem instead of giving them time to think independently).

The goal of their involvement must be to improve learning, not test scores. They can help by respecting their children's learning style and space: some prefer absolute silence, others like to listen to soft music; some like to daydream or watch

YouTube for a while and then intensely focus on their homework, others prefer to do their homework without any distraction; some like to concentrate on one task, others like to multitask (research doesn't say never to multitask; advise your children not to multitask while they are trying to learn something new that they hope to remember). And they should forget about punishing their children for not doing their homework. This strategy never works and at times it backfires. They should let teachers intervene if the child isn't doing homework correctly or regularly. Meeting with teachers to discuss their child's low scores doesn't really help the child do well. It just makes the child more anxious.

Parents should provide support in the form of prompts, hints, suggestions and reminders rather than doing their children's homework. Research shows that parental support for autonomy instead of parent rule setting is associated with higher standardised test scores, higher class grades and more homework completed. Laurence Steinberg, Temple University psychology professor and author of *The Ten Best Basic Principles of Parenting*, advises parents not to help with homework unless:

- the child asks for something specific that is beyond child's capacity;
- the child doesn't understand what the homework assignment is (and parents can explain it concisely and confidently); or

- the teacher has explicitly set an assignment that requires the parent and child to work together.

Overprotective parents not only help with homework, they also make sure that they pick their children's university courses. Studies have found that more parents help children select a college course, the less satisfied the student is with that course. So much so for helicopter parents' dream of seeing their child wearing a stethoscope around their neck and examining a patient or wearing a barrister wig and cross-examining a criminal.

In her best-selling book, *All Joy and No Fun: The Paradox of Modern Parenthood*, American journalist Jennifer Senior says that parents are making themselves miserable by believing they always have to maximise their children's success. Below is an excerpt from a scene she describes from 1540 hours of video footage—collected by University of California researchers—documenting day-to-day lives of 32 middle-class families which captured all the fights between couples, all the negotiations between parents and their children. In this scene, the mother is approaching her eight-year-old son, the oldest of two, who's seated at the computer in the den, absorbed in a movie. At issue is his homework, which he still hasn't done. Mother and son are arguing—tensely, angrily—and she is pulling on his arm.

The boy reaches for the keyboard. 'I'm putting it on pause!'

'I want you to do your homework,' his mother repeats. 'You are not –'

'I know,' the son whines. 'I'm going to pause it!'

His mother's not buying it. What she sees is him stalling. She pulls him off the chair.

'No, you're *not*,' says his mother. 'You're still not listening!'

'Yes I am!'

'No, you're not!'

[The tense argument continues for a long while and the son is still watching the movie when the recorded scene ends.]

In my opinion, the mother would be better off if she did her son's homework like James Ellroy's father. The son may turn out to be a famous novelist. A win-win situation for both mother and son. Just kidding.

Chapter 12

Let children daydream

learn daydreaming helps children to explore new ideas/*unlearn* daydreams are the delusions of the devil

> A certain amount of daydreaming is good, like a narcotic in discreet doses. It lulls to sleep the fevers of the mind at labour, which are sometimes severe, and produces in the spirit a soft and fresh vapour which corrects the over-harsh contours of pure thought, fills in gaps here and there, binds together and rounds off angles of the ideas. But too much daydreaming sinks and drowns.
>
> Victor Hugo, *Les Misérables*

An empty mind is the devil's workshop, so goes the old saying. But our minds are never empty; neurons are always hard at work even when we are sleeping, dreaming or daydreaming. The brain is an expensive organ to run; it consumes about 20 per cent of our daily calorie intake. If we're not engaged in a task, it makes no sense that our brains would just turn off.

We daydream for about one-third of our waking hours, although a single daydream lasts only a few minutes. The reason we spend so much time daydreaming because the human brain has been purposefully designed for daydreaming; it is our minds' default mode of thought. Brain scans show that our brains have a 'default network', interconnected brain regions, which remain active when we daydream or let our mind wander.

This default network functions more vigorously when the brain has no specific task to focus on. When we have a pressing task, the brain focuses on that task and the default network is relatively suppressed. But during familiar tasks, such as doing the dishes when there is no external demand for thought, our minds do not go blank. Instead they tend to wander, moving swiftly from one thought to the next, generating images, voices and feelings. Most of the time the wandering thoughts are not fanciful; they usually are personal thoughts like working out everyday problems or making plans for the near future.

Undoubtedly, as you are reading this page your mind has already drifted off to thoughts unrelated to what you're reading. When your mind wandered off, it was still using resources of the brain. Consider what happens when you try to remember a telephone number or email address while you are looking for a pen and paper to write it down. This information is in your working memory, which is a kind of mental workspace, where we have the ability to retain and mentally manipulate information over short periods. Working memory capacity has been correlated with general measures of intelligence such as reading comprehension and IQ scores. New research now shows that working memory enables the maintenance of mind wandering.

People with high working memory tend to daydream more. Daydreaming is an indication of underlying priorities being held in the working memory. 'But doesn't mean that people with high working memory capacity are doomed to straying mind,' says psychologist Daniel Levinson of the University of Wisconsin. 'The bottom line is that working memory is a resource and it's all about how you use it. If your priority is to keep attention on task, you can use working memory to do that, too.' Children, who are natural and prolific daydreamers, can daydream as well as focus on the task.

An online parenting magazine advises: 'Daydreaming is a behavioural disorder. Daydreamers are actually not in touch with the reality ... You should try and curb your child's daydreaming tendencies at as young an age as possible.' These extreme thoughts seem to have come from some misguided Tiger-Mom-type parent. But most parents and teachers can be accused of rousing children out of their reverie and scolding them to focus on the task in front of them.

There is extensive evidence that daydreaming is not a waste of time. It helps children to make meaning out of experience and information they encounter. It helps make children creative and improves their school performance. Imaginary friends benefit children's language skills. Imaginary scenarios and make-believe games help them in understanding complex emotions and social skills. Daydreaming is relaxation, a kind of micro-holiday from which children come back fully recharged and refreshed. Children who do not daydream enough (because they are too busy watching television) tend to be unimaginative.

Jerome Bruncer, an American psychologist who in the 1980s was involved in a fascinating project 'Narrative from the Crib', recorded a two-year-old girl's conversation she had with herself before she fell asleep. What he found was that her conversations with herself were significantly more advanced than her conversations with her parents. 'She would create a story to try to integrate events, actions, and feelings into one structure—a process that is a critical part of a child's mental development,' writes Malcolm Gladwell in *Tipping Point: How Little Things Can Make a Difference*.

Research by Mary Helen Immordino-Yang, an education professor at University of Southern California, and her colleagues suggests that parents and teachers should be encouraged to teach children the value of more diffuse mental activity that characterises our lives: daydreaming, remembering and reflecting. The researchers explain that the brain has two operating systems: 'looking out' directs out attention to get the things done; 'looking in' directs us inwards, setting our thoughts. When our brain is engaged in 'looking in' mode it makes sense out of experiences and information it encounters when we are 'looking out'.

Schools demand constant attentiveness and a hyper-connected world ruled by social media draws attention away from the world inside. There is little time left for daydreaming. Ironically, it diminishes children's capacity to pay attention when they need to. 'If youths overuse social media, if they spend very little waking time free from the possibility that a text will interrupt them,' the researchers say, 'we would expect that these conditions might predispose youths toward focusing on the concrete, physical and immediate aspects of situation and self, with less inclination toward considering the abstract, longer-term, moral and emotional implications of their and other's actions.'

The daydreaming mind is not shackled to its immediate surroundings; it is free to go anywhere; it's free to make new associations and connections; it's free to engage in abstract thought and imaginative ramblings. The intense focus on a problem has its advantages, but the relaxed style of thinking leads children to contemplate ideas that sometimes seem silly or far-fetched. Such imaginative thoughts might not be practical, but they often are the perfect springboard to creative insights. Children, like everyone else, need their daydreams, as Victor Hugo says, to round off angles on their ideas.

How should parents and teachers respond to the benefits of daydreaming? 'For one thing, we should stop snapping our children out of their daydreams,' urges Jessica Lahey, a former teacher and author of *The Gift of Failure: How the Best Parents Learn to Let Go So Their Children Can Succeed.* 'Instead, we should protect this time much as we protect bedtime.'

Parents and teachers must also come out of the era when daydreaming was considered idleness. By all accounts, it's an industrious occupation.

> 'What teachers and the administration in that era never seemed to see was that the mental work of what they called daydreaming often required more effort and concentration than it would have taken simply to listen in class.'
>
> David Foster Wallace, *Oblivion*

Chapter 13

The myth of superior children

learn science doesn't support 'pushy parenting'/*unlearn* children in some cultures are superior to others

> 'It concerned a scientist called James Morningdale, quite talented in his way ... What he wanted was to offer people the possibility of having children with enhanced characteristics. Superior intelligence, superior athleticism, that sort of thing ... Children demonstrably *superior* to the rest of us?'
>
> – Miss Emily to Kathy
> Kazuo Ishiguro, *Never Let Me Go*

Mums and dads who want to parent well struggle to find the right balance between encouragement and discipline. Too much discipline makes parents 'authoritarian'; they rely on punishment—or threat of punishment—to control their children. At times, punishment also extends to withdrawal of parental affection. Such parents stifle any attempt by their children at independence or autonomy. Children from authoritarian households are more compliant but studies show that they suffer from anxiety, depression and poor self-esteem.

Psychologists associate 'authoritative' parenting with best outcomes for children. The strictness of these parents is reasoned and reasonable and is accompanied by warmth and encouragement of self-direction. Authoritative parents do a lot of explaining of their rules and are responsive to a child's changing needs. They also adapt their parenting as a child enters a new phase, a quality that is particular important in teenage years, says American psychologist Laurence Steinberg, one of the foremost researchers on parenting style. Children raised in such households are generally well behaved, socially adept and psychologically healthy.

Parenting style is a huge influence on a child but it's only an influence. The child's peer groups have an influence too. As far as success in school is concerned, children from both authoritarian and authoritative households do well. A recent study by Steinberg of more than 200,000 students in US schools from all ethnic backgrounds also confirmed that that children raised in authoritarian households got grades comparable to children from authoritative households.

This wisdom contradicts assertions made by Amy Chua in her best-selling book, *Battle Hymn of the Tiger Mom*, that successful parenting demands controlling most aspects of a child's life. 'What Chinese parents understand is that nothing is fun until you're good at it,' she writes. 'To get good at anything you have to work and children on their own never want to work, which is why it is crucial to override their preferences.' She never allowed her two daughters to have play dates, sleepovers, watch TV or choose their own extracurricular activities such as taking part in school plays. She screamed at her daughters when their grades were less than an A, and admits to insulting them by calling them 'garbage' and 'fatty'.

To Chua, enjoyment, interest and freedom of choice are somehow incompatible with hard work, persistence and success, says Heidi Grant Halvorson, author of *Succeed: How We Can Reach Our Goals*. 'Again and again, research has shown that when children feel they have choices, it creates *intrinsic motivation*—the desire to do something for its own sake,' Halvorson says. 'With choice, they enjoy what they are doing more. They are more creative, process information more deeply, persist longer and achieve more. Intrinsic motivation is in fact *awesome* in its power to get and keep us going. Your kids will work hard of their own free will, and even have fun doing it, when you *don't* completely override their preferences.'

Chua claims that aggressive parenting is good and that's how most Asian mothers parent. But, in reality, no study on parenting supports either of her claims. By helping children to make decisions for themselves parents allow children to bounce back from disappointment. By coercing them into compliance, parents only create psychosomatic problems for their children.

The famous Tiger Mom's parenting style may be 'superior', at least according to her own calculations, but when she applies her idea of 'superiority' to success and achievement of certain groups in America it seems like old distasteful racism in new kitschy clothes. *The Triple Package: How Three Unlikely Traits Explain the Rise and Fall of Cultural Groups in America*, written with her husband Jeb Rubenfeld, argues that not just Chinese mothers but Jewish, Indian, Iranian, Lebanese, Nigerian, Cuban and Mormon parents are superior to their American counterparts.

And the success of their children rests on three pillars: (1) a superiority complex—a deep-seated belief in their exceptionality (in the case of Jews and Mormons it emanates from their religious doctrine and makes them outperform those of other groups); (2) insecurity—a feeling that you or what you've done is not good enough (a triple-package family would not hesitate in shaming their children if they scored C or B in a test); and (3) impulse control—essentially self-discipline which makes them persevere through adversity (Mormon missionaries, the book points out, are always getting doors slammed in their faces). Immigrants in every country are an ambitious bunch and many successful immigrants have these three traits; but the authors fail to explain how these traits lead to better outcomes.

Like any other book on success, it cherry-picks success stories. It's over represented by high-earning business people or high-achieving academics. Its recipe for success only applies to those who are already successful. Its definition of success hovers somewhere in the centre of money, fame and power. It's not everyone's idea of success. 'For all their struggling, the children of immigrants end up earning the same, on average, as other Americans do,' remarks Olga Khazan in *The Atlantic Weekly*, an e-magazine.

Occasionally, my local Indian community newspapers carry stories showing off the 'superiority of being an Indian' by quoting questionable facts and figures such as 38 per cent of doctors and 12 per cent of scientists in the US are Indians and zero, decimal, point, algebra, quadratic equations, trigonometry and calculus were invented in India. *The Triple Package* is a book-length version of the superiority-of-being-an-Indian kind of stories I read in my community newspapers or on the internet. 'Well, if Indians are so great, what explains India?' asks Suketu Mehta, the New York-based author of *Maximum City: Bombay Lost and Found.* 'The country is a sorry mess, with the largest population of poor, sick and illiterate people in the world, its economy is diving, its politics abysmally corrupt.'

On balance, *The Triple Package* seems like an empty package. An old-fashioned phrase—empty vessels make the most noise—aptly describes the book's old-fashioned idea of superior ethnic groups.

Chapter 14

Learn to learn

learn by reflecting, self-testing, practising and thinking like a child/
unlearn the only way to learn is to work harder

'If we learn why we do things, we won't make the same mistakes.'

– Ann to Olive
Elizabeth Strout, *Olive Kitteridge*

The old dogma that the human brain fossilises as we age and by the time we reach adulthood it pretty much follows the old proverb 'you can't teach an old dog new trick' has now been overthrown by the new idea of neuroplasticity that the brain can be rewired. We have some control over how the brain rewires itself: learning new skills. Learning involves strengthening connections between neurons. This can happen in two ways: by creating more connections between neurons, and by increasing their ability to send signals.

At times learning can be as simple as thinking, as shown in a classic experiment by American neuroscientist Alvaro Pascual-Leone. He compared the brain scans of two groups of volunteers: those who practised piano exercises for a week, and those who merely thought about practising the piano exercises (holding their hands still while imagining how they would move their fingers). Connections between the neurons in the brains of both groups reorganised in a similar way. Seems strange that even ephemeral thinking can alter the form and function of enduring brain.

Learning is more effective when we take out time to deliberately focus on thinking about what we have been doing. If we'd take some time out for reflection, we might be better off, says a team of researcher led by Giada Di Stefano of HEC Paris who studied the effect of reflection on learning. By reflection, the researchers mean taking time out after a lesson to synthesise, abstract and articulate the main points. Reflection, they add, builds our confidence in the ability to achieve a goal. The researchers' own 'reflection' comes from a study based on dual-process theory of thought which says that we think and learn using two distinct processes: type 1 processes are fast, automatic and intuitive (learning by doing); and type 2 processes are slow, conscious and controlled (reflective and associated with decision making).

In the laboratory part of the study 202 adults completed an online maths brain-teaser under three different conditions: *reflection* (participants took a few minutes to write what strategy they used or might use in the future to solve problems before starting the second round); *sharing* (participants were told that their notes will be shared with future participants); and *control* (participants simply completed another round of brain-teasers). Participants who were allowed to reflect performed an average of 18 per cent better on the second round than the participants who were not given time to reflect. The field study was conducted at a tech support call centre where newly hired customer service employees undergoing one-month training were compared. Again, those given time to think and reflect scored 23 per cent better on their assessment than those who were not.

Stefano and colleagues conclude that their research confirms American philosopher and educational reformer John Dewey's words: 'We do not learn from experience ... we learn from reflecting on experience'. The message is clear: stop, reflect and think about learning.

The idea of mindset (*see* the next story 'Fearful of failure?') shows that by changing mindset in a certain way we can accelerate learning. Gabriele Wulf, an American expert on motor performance and learning, has studied the effect of mindset on tasks that involve perceptual learning or motor skills such as coping with tinnitus or perfecting a golf swing. In a study two groups of volunteer golfers were given different sets of instructions: focus on the movement of the arm (internally on your body movement) or focus on the club movement (externally on the resulting action). The external group scored twice as high as the internal group. By simply shifting focus during sporting activities we can improve results.

In another study to determine the effect of altering mindset on learning a novel physical exercise, Wulf provided older women (average age 71 years) false feedback that their performance was above average. This reduced concerns about their learning ability and nervousness, resulted in more effective learning. In the second experiment older women (average age 64 years) were told that their peers usually do well on the task. This also enhanced their learning. The results demonstrate that motor performance and learning in older age can be influenced quickly and positively by having a positive motivation or mindset.

Children learn faster because they throw themselves into the task. Adults worry more about whether they're doing things right. 'Adults think so much more about what they are doing,' Wulf says. 'Children just copy what they see'. To learn faster, learn like a child.

You have heard the old saying 'practice makes perfect', but now scientists have found the evidence to support how practice rewires the brain, at least in laboratory mice. Developmental biologist Yi Zuo of the University California and her colleagues studied mice as they learned new behaviours such as reaching through a slot to get a seed. After each day's training the researchers observed changes in the brain layer that controls muscle movements during training. Mice that repeatedly practised for four days grew new structures on their neurons. These cellular structures called 'dendrite spines' form synapses that connect neurons. One-third of the newly formed spines were located next to another new spine. Neurons work very hard to form clusters, to place spines close to one another. These clusters magnify the strength of connections between neurons. The clusters in the brains of mice were specific to the task being learned. The durable connections between a group of neurons form the memory of a new task learned. But practice is of no help if it's not perfect; that's it is focused on quality. Perfect practice makes perfect, indeed.

Like practice, self-testing also improves learning by strengthening memory. During the past hundred years, hundreds of experiments have shown that testing improves learning. A fact known even to Aristotle more than 2,300 years ago: 'Exercise in repeatedly recalling a thing strengthens the memory'. But until now no one was sure why. Psychology researcher Katherine Rawson of Kent State University in the US and her former graduate student Mary Pyc asked 118-English-speaking students to learn 48 Swahili words by pairing them with their English counterparts (for example, *wingu*-cloud, *lulu*-pearl and *zabibu*-grape, if you want to self-test yourself after a few minutes). They hypothesised that learners invent a mediator—word, phrase or concept that connects one piece of information to another—to trigger the right piece of information. For example, 'wing' might serve as a mediator between *wingu* and cloud (a bird flying in the clouds).

When the students were tested one week later, as expected

those who had taken the practice test did better than those hadn't taken the test but only studied. When Rawson and Pyc asked students to recall their mediators just before the test, those who had taken the practice test remembered their mediators 51 per cent of the time. Those who hadn't taken the practice test remembered their mediators only 34 per cent of the time. 'The illusion is, you read something and think you'll remember it,' says Pyc. 'But if don't try to retrieve it, you don't know if you know it.' Time for a test: Repeat the three Swahili words you learned a few minutes ago.

The pen is mightier than sword—and the keyboard. During lectures, taking notes on laptops rather than longhand is increasingly common. Pam Mueller, a psychology graduate student, had left her laptop and was forced to use pen and paper to take notes in a psychology lecture. Surprisingly, she felt like she had gotten much more out of the lecture that day. The inkling that there might be something different about writing led Mueller, now at Princeton University, to collaborate with Daniel Oppenheimer of the University of California, the professor teaching the class. Their research suggests that 'even when laptops are used solely to take notes, they may still be impairing learning because of their use results in shallower processing'. When you take notes by hand, you process information as well as write it down, but when you are typing you tend to just transcribe large chunks of lectures verbatim. That initial selectivity leads to long-term comprehension, says Mueller.

Whether you take notes by laptops or by longhand, traditional lecture classes are ineffective and undergraduate students in sage-on-a-stage-type lecturers are one and half times more likely to fail than students in classes that engage students with questions or group activities. This is the conclusion of biologist Scott Freeman of the University of Washington and group of colleagues after analysing 225 studies of undergraduate teaching methods in science, technology engineering and maths.

Other learning techniques that don't work are highlighting and rereading. Highlighting is simple and quick but it doesn't improve learning. There is also no clear evidence that rereading improves comprehension. The best strategy is to spend time on self-explanation and self-testing.

We all make mistakes but when we are aware of our mistakes we can easily correct our behaviour. People who think they can learn from mistakes have brains that are more tuned to pick up on mistakes very quickly. Their brains are tuned differently at a very fundamental level. What you need is the right mindset: growth-mindset is not simply dependent on intelligence; motivation and effort are huge part of it. 'A growth mind-set is about focusing on the process—as in the experience—rather than only on the outcome,' says psychologist Jason Moser of Michigan State University. When growth-minded people make mistakes they pay attention to what went wrong and get the information they need to improve. On the other hand, people who think that they can't get smarter—fixed-mindset people—will not take opportunities to learn from their mistakes.

Learning is a 'beautiful reciprocal arrangement', as Mr Vinson, a teacher, tells to Holden Caulfield, the teenage protagonist of J. D. Salinger's *The Catcher in the Rye*: 'You'll learn from them—if you want to. Just as someday, if you have something to offer, someone will learn something from you. It's a beautiful reciprocal arrangement. And it isn't education.'

Chapter 15

Fearful of failure?

learn not to let fear of failure stop you/*unlearn* that your talent is fixed from birth

> 'There is only one thing that makes a dream impossible to achieve: the fear of failure.'
>
> – the alchemist to Santiago, a shepherd boy, when he says, 'I have no idea how to turn myself into the wind.'
>
> Paulo Coelho, *The Alchemist*

Once seeded, fear of failure sprouts in the brain making it incapable of making decisions. This stagnation ensures that we only meet failure. When I started researching this story I found gazillions of papers on fear of failure churned out by 'fearful' psychologists, or cognitive scientists as some prefer this label. It's not for me but for some aspiring young psychologist to find out why so many of their colleagues are preoccupied with failure. I looked at my desk bulging with books, research papers and pop articles on fear of failure; and before fear of failure could grip my brain, I started writing following this astute advice: failure fears people who pursue meaningful goals.

I'll start with the work of Carol Dweck, an eminent psychologist and author of *Mindset: The New Psychology of Success*, who has spent decades studying how people cope with failure. She came up with the idea of mindset when she was sitting in her office studying the result of the latest experiment with one of her graduate students. The results showed that people who disliked challenges thought that talent was a fixed thing that you were either born with or not. People who relished challenges thought that talent was something you could nourish by doing things you were not good at all. 'There was this eureka moment,' recalls Dweck.

She later came up with the terms 'fixed mindset' to identify the former group and 'growth mindset' for the latter group. If you believe you can develop your talents over time (a growth mindset), you'll never be paralysed by fear of failure. If you believe you were born with a certain amount of talent (a fixed mindset), that's the end of the road for you. A growth mindset benefits us throughout our life. 'It allows you to take more challenges,' she says, 'and you don't get discouraged by setbacks or find effort undermining.' She applies her research findings to her own life. She took up piano as an adult and learned Italian in her 50s. 'These are things the adults are not supposed be good at learning,' she says. 'Just being aware of the growth mindset, and studying it and writing about it, I feel compelled to live it and benefit from it.'

We can learn to change our mindsets and make dramatic stride in our performance. 'Changing mindsets is not likely surgery,' Dweck warns, 'You can't simply remove the fixed mindset and replace it with growth mindset.' First, you have to learn that talent is like a muscle, which grows stronger through exercise, and then train yourself to master new things. Proverbial practice may not make you perfect but it will certainly improve your performance.

Are you burdened with fear of failure? Write *true* or *false* against the following statements whether they are generally like you or not. This is not a diagnostic test; it may help you in finding out the areas you need to work to change.

1. Failure makes me worried what other people would think about me.
2. I'm afraid of looking dumb.
3. I'm uncertain about my ability to avoid failure.
4. I like to play it safe as I can't afford to be vulnerable.
5. I always put off tasks for tomorrow.
6. I become anxious when not certain.
7. I live in self-doubt.
8. I'm afraid of disapproval.
9. I worry that I won't do well.
10. I worry that failure would disappoint people whose opinion I value.

Any *true* answer suggests that you might like to examine the issue further. But don't be too hard on yourself. Failure to act correctly is an inevitable part of life; that's why computer keyboards have delete keys. You can always delete a failure from your memory and start again. It's easier if you have a growth mindset: every failure will rewire the brain making it stronger to face the next big challenge.

Here're some of the ways to lose your fear of failure:

Maintain perspective. Take a long-term view of your failures; they are not final. A failure is a single incident; it doesn't make you incapable of success in the future. Failure is a relative term. Was Vincent von Gogh's inability to sell more than one painting in his lifetime a failure? If he had fear of failure humanity would have been deprived of the eternal beauty of his nearly 2000 paintings, drawings and sketches.

Think of failure as a learning experience. Put aside old ideas and past efforts and start anew. Visualise your goals; workout your milestones. Develop a strategy—a step-by-step plan that makes sure your actions lead you towards your objective—and execute it efficiently. Let Thomas Alva Edison inspire you: after experimenting with thousands of different sorts of fibres (including the hair from the beards of some of the men in his laboratory) he at last found the right filament for his newly invented incandescent light bulb. He hadn't failed thousands of times; he had found thousands of ways that didn't work. The one that worked brought sunshine into our darkened rooms.

Identify things that are in your control and focus on them. You may not be good at figures (of mathematical type), but if you like to draw you can focus on figures (of curvaceous type). Everyone has talents they are not sure about or even do not know about. Once you know of your unique gift, you know of one thing that is under your control. Just focus on it.

Failure is not defeat and success is not excellence. Plunge right into what you really want to do. It's better to enjoy partial success than nursing regrets of not doing it at all.

To avoid emotional bruises caused by failure, learn to own the fear. Find trusted people with whom you can discuss your demoralising feelings of shame and disappointment. 'Bringing these feelings to the surface can help prevent you from expressing them via unconscious efforts to sabotage yourself; and getting reassurance and empathy trusted from others can bolster your feelings of self-worth and minimise the threat of disappointing them,' advises psychologist Guy Winch, author of *Emotional First Aid: Practical Strategies for Treating Failure, Rejection, Guilt, and Other Everyday Psychological Injuries*.

Parents are keen to motivate their children, but fear of failure is not the right tool to motivate them. Many parents believe that if you want to develop you children's resilience, you let them fail and don't hide their failure from them. The argument is that when children don't get a hoped-for reward, they will be motivated to try harder next time. Negative motivation is educationally and psychologically weak. Alfie Kohn, author of *The Myth of the Spoiled Child: Challenging the Conventional Wisdom about Children and Parenting*, dismisses let- children-fail style of parenting. 'When kids' performance slides, when they lose enthusiasm for what they're doing, or when they cut corners, much more is going on than laziness or lack of motivation,' she writes. 'What's relevant is what their experiences have been. And the experience of having failed is a uniquely poor bet for anyone who wants to maximise the probability of future success.'

A study of 347 UK students, average age 15, over an 18-month period also confirms that failure-focused messages make students less motivated to do well. The students received both negative and positive messages from their teachers before the exam. For example, 'If you fail the exam, you will never be able to get a good job or go to college. You need to work hard in order to avoid failure.' Or, 'The exam is really important as most jobs that pay well require that you pass and if you want to go to college you will need to pass the exam.'

'Both messages highlight to students the importance of effort and provide a reason for striving,' says researcher David Putwain of Edge Hill University in England. 'Where these messages differ is some focus on the possibility of success while others stress the need to avoid failure.' The study results showed that students who felt threatened by teachers' failure-focused messages felt less motivated to do well and had lower exam scores than those whose teachers used fewer fear tactics.

Forget fear tactics; try this tactic, not from a psychologist but from an acclaimed writer of short fiction:

> When we begin to take our failures non-seriously, it means we are ceasing to be afraid of them.
>
> Katherine Mansfield in 'A Shot of Laughter'

Chapter 16

Your secrets revealed

learn secrets are suppressed thoughts and keeping them requires constant effort/*unlearn* you can keep a secret forever; as Freud said if your lips are silent, betrayal oozes out of your every pore

> 'Mae, have you ever had a secret that festered within you, and once that secret was out, you felt better?' [Bailey asked]
> 'Sure.'
> 'Me too. That's the nature of secrets. They're cancerous when kept within us, but harmless when they're out in the world.'
> 'So you're saying there should be no secrets.'
> 'I have thought on this for years, and I have yet to conjure a scenario where a secret does more good than harm. Secrets are the enablers of antisocial, immoral and destructive behavior.'
>
> Dave Eggers, *The Circle*

When Leo Tolstoy was five one of his older brothers told him to stand in the corner until he stopped thinking about the white bear. It seems a simple enough command but Tolstoy was unable to do it. Instead, his mind was seized by fear of the unwanted thought and he thought of nothing else but white bears.

This story of fear of white bears inspired American psychologist Daniel Wegner in 1987 to conduct an experiment in which he gave participants a white-bear task to test the effectiveness of suppression of unwanted thoughts. Participants were asked to enter a room alone with a tape recorder and record everything that came to mind for five minutes. Before the experiment, Wegner told some participants to think of a white bear, and told others not to think of a white bear. All participants were to ring a bell if they merely thought of a white bear. Like little Tolstoy, almost all participants kept thinking again and again of that banned white bear and rang the bell once a minute.

When Wegner asked participants to switch roles, he found that those participants who were originally told not to think of white bear but now allowed to think of white bear (they were now free to express a banished thought) rang the bell significantly more times than the other group. The suppressed thought gushed out with greater frequency than it had not been suppressed earlier. He called this the 'rebound effect': as we relax our efforts to suppress, we experience a resurgence of the suppressed thoughts.

'By suppressing a thought, we never get used to it,' says Wegner whose 1989 book *White Bears and Other Unwanted Thoughts* is now a psychology classic. 'Instead, we make ourselves more sensitive to its next occurrence. The thought we most want to avoid soon becomes our greatest fear.' To end obsessive thoughts, he advises, stop stopping them.

To set the white bears free, accept unwanted thoughts instead of suppressing them; and disclose your problems rather than keeping them secret. We can suppress thoughts only for a limited time and the efforts to suppress become less successful over time. They will come out eventually.

Why is it so difficult to suppress unwanted thoughts? The answer lies in the prefrontal cortex, the part of the brain right behind the forehead that is responsible for planning and mental control. By overburdening the prefrontal cortex with suppressed thoughts and secrets, we compromise its capacity of making decisions. In an experiment, participants who watched a video while trying to ignore words that flashed on the screen—an act of self-control that taxed the prefrontal cortex—performed less well at subsequent self-control tasks than participants who drank a glucose drink after performing the first task. Glucose replenished the fuel brain needs to function. Suppressing thoughts depletes glucose, which later reduces the ability to suppress. Nice idea, but having a glass of glucose drink daily may not help you forget your old flame or keep a dirty secret from the new one.

At times suppressed thoughts find their way into our dreams. The prefrontal cortex is less active during rapid-eye-movement (REM) stage of sleep, which diminishes the brain's ability to keep suppressed thoughts at bay. 'Maybe this is why students dream of sleeping through an important exam, why actors dream of going blank on stage, and why truckers dream of driving off the road,' Wegner suggests. 'Dreams are where our thoughts go when we try to put the thoughts out of mind.'

A team of American researchers led by Michael Slepian of Stanford University suggests that keeping secrets exerts our mental abilities and this mental exertion might actually wear the body down. In one of their experiments, they examined the suppression of a commonly kept secret: sexual orientation. After answering questions in front of a video camera and completing a questionnaire about sex, age, ethnicity, sexual orientation and personality, the participants were requested to move stacks of books as the lab was ostensibly relocating. The participants concealing their sexual orientation lifted fewer stacks. From this and other experiments the researchers conclude that important meaningful secrets, including those regarding infidelity and sexual orientation, affect people in various ways, as if they were physically burdened.

Keeping an emotionally charged secret not only physically burdens the body; it also causes ailments ranging from colds to chronic diseases. The findings of a recent survey of 790 Dutch teenagers show that teens who confide in a parent or close friends report fewer physical complaints and less delinquent behaviour, loneliness, lower quality relationship and depression than those who hide their secrets. The findings prove the old adage that secrecy is bad while sharing is good.

People who do not feel like disclosing upsetting or traumatic experience of their lives (such as abuse, alcoholism, divorce, loss of loved ones, suicide attempts etc.) are advised to write about them. The preponderance of evidence suggests that putting experiences into words has powerful effects. People who write about stressful life experiences for a few minutes each day for a few consecutive days show surprisingly beneficial health effects that can last for months. For some patients it even serves as a useful supplement to regular medication.

Following on the idea of a 'writing cure' numerous studies have now demonstrated that disclosing a personal secret—from telling someone to writing it on a piece of paper that is later burned—boosts both physical and mental health. On the flip side, those who are secretive tend to be more shy, anxious and depressed, which makes them more vulnerable to illness.

Deciding to reveal a secret that is creating an emotional barrier between you and someone close to you is hard, indeed. But you can make it less hurtful for the person who would be hurt more by learning it from someone else. By telling the secret you may even jeopardise your relationship, but by keeping it within you will only turn it into a cancerous thought. Think of it this way: when you do something that would hurt someone, you have committed the first 'offense'. The second 'offence' is keeping it from them. Would you like to plead guilty of one charge or two charges?

Secrets are never sweet; they are miseries of mind. They take up more space in the brain when we try not to think about them. Why give that useful space to white bears?

Chapter 17

Prejudices polluting our unconscious minds

learn to admit, confront and reduce prejudice/*unlearn*
'All nice people, like Us, are We/And everyone else is They'
(Rudyard Kipling)

'And never allow yourself to be blinded by prejudice?'
'I hope not.'
'It is particularly incumbent on those who never change their opinion, to be secure of judging properly at first.'

— Elizabeth Bennett to Mr Darcy
Jane Austen, *Pride and Prejudice*

Like weeds, prejudice is a word laden with unpleasantness. In *The Devil's Dictionary*, published in 1906, American journalist Ambrose Bierce defined it as 'a vagrant opinion without visible means of support'. Even after more than a century this definition still makes sense. All negative stereotypes of age, gender, race, religion, skin colour, ethnicity, social class, sexuality, disability status, nationality and so on are vagrant opinions without any basis in reason or research.

Like it or not, we all are prejudiced. We may explicitly identify ourselves as unbiased but unconsciously we make lazy assumptions about people based on whether they are men or women, black or white, fat or skinny. But no one would admit to being prejudiced. We immediately know when we encounter prejudice but we do not know that we all unconsciously harbour prejudices, which we consciously reject. These hidden or implicit prejudices occur outside of our consciousness awareness and control. For example, you may believe that boys and girls are equally good at maths but it is possible that without knowing it you could associate boys with maths more than you associate girls with maths. We would say that you have an implicit boys-are-better-at-maths prejudice. Implicit prejudices are more prevalent than more explicit prejudices, which we associate with slavery or Nazism.

Interested in digging out prejudices that are buried deep within your mind? Google 'Project Implicit' and then spend a few minutes to take a popular online test—implicit association test or IAT—to find out implicit attitudes that you did not know about. The IAT is free and have been taken by millions of people around the world since its start in 1988.

The IAT is the brainchild of leading American social psychologists, Anthony Greenwald and Mahzarin Banaji, who have spent decades studying implicit social cognition, the thoughts and feelings outside of consciousness awareness and control. When Greenwald created the first IAT in 1988 he measured how quickly people tapped keys on a computer keyboard in response to a prompt on the screen. He found, as he had predicted, that people more easily associated positive words such 'happy' or 'peace' with pictures of flowers and negative words such as 'rotten' or 'ugly' with insects. When he began testing responses to words and images associated with race and ethnicity, participants' automatic reactions didn't match the attitude they said they had. He had discovered a tool to investigate our hidden biases.

The IAT gives you an opportunity to delve into conscious and unconscious biases for more than 90 different topics ranging from pet to political issues, ethnic groups to sports teams, and entertainers to style of music. The IAT is designed to measures the strengths of associations between concepts (for example, white people, black people) and evaluation (good, bad). The score is based on how long it takes, on average, to press the corresponding keys. A white test-taker would reveal underlying prejudice if the response is faster for black/bad than black/good. Many people who consider themselves egalitarian and believe they treat all people fairly and equally are nonetheless surprised by the greater difficulty suggested by a slower pressing of black/good keys.

The IAT data has revealed some interesting insights. The race IAT tells that 75 per cent of its takers in the US, including some African Americans, have an implicit preference for white people over black people. The results of body weight test are equally disturbing: they show an overwhelming automatic preference around the world for thin people over fat people.

'We really do believe that we pretty much know what goes in our heads,' says Banaji's, a long-time collaborator of Greenwald. 'And that's because we do have access to a piece of it called the conscious mind, and that wrongly gives us the feeling that we know all of it.' When Banaji herself sat down to take the test for biases against women, names of men and women and words associated with 'career' and 'family' flashed across the computer screen, one after another. As she tried to sort the words into groups as asked, she found she was much faster and more accurate when asked to group the male names with job-related words. You wouldn't expect a trail-blazing woman social psychologist to be biased. No one is immune from hidden biases, including the test designers. Banaji insists that the IAT is not designed to shame people. Its aim is to make us aware of blind spots in our unconscious mind. But, fortunately, our unconscious minds are not permanently stuck on prejudices and stereotypes. By weaving awareness into our day, says Banaji, we can help our conscious attitudes take change.

Where do prejudices come from? Arne Roets and Alain Van Hiel of Ghent University in Belgium say that they do not come from ideology, but rather a basic human need and way of thinking. People who are prejudiced feel a much stronger need to make quick judgments and decisions. They hate uncertainty and therefore quickly rely on the most obvious information, often the first information they come across, to reduce it. That's also why they favour authorities and social norms which make it easier to make decisions. Then, once they've made up their mind, they stick to it. If you provide information that contradicts their decision, they just ignore it. This way of thinking is linked to people's need to categorise the world, often unconsciously. 'When we meet someone, we immediately see that person as being male or female, young or old, black or white, without really being aware of this categorisation,' Roets says. 'Social categories are useful to reduce complexity, but the problem is that we also assign some properties to these categories. This can lead to prejudice and stereotyping.'

These days it seems no research on the brain is complete without mentioning the hormone oxytocin, variously described as a 'compassionate hormone', 'trust hormone', 'cuddle hormone, 'love hormone' and even 'moral hormone'. It has now earned a new label, 'prejudice hormone'. Psychologists at the University of Amsterdam have found that Dutch men who inhaled oxytocin were more likely to associate positive words such as joy and laughter and complex positive emotions such as hope and admiration with Dutch people than with Germans and Arabs.

The best way to fight prejudice is—not by avoiding oxytocin—but by admitting prejudice. We are not born prejudiced; we pick it up as we grow from family, friends, books, movies, television etc. Our prejudices can be modified and can be changed if we are motivated to think about someone, in some way, as a member of our own group. We all are one people.

Participation in cultural activities of another racial or ethnic group also helps in reducing prejudice. A team of Stanford University psychologists led by Tiffany Brannon have found that people's attitudes towards another racial or ethnic group improve when they participate in the other group's cultural activities, even as simple as an activity as making a music video together. 'What's empowering about this research is that too often people feel they need to hide their cultures,' the researchers say, 'but this suggests bringing it out, having people being engaged in it, can improve attitudes and make more pleasant environment on the whole.' Other studies also show that more meaningful face-to-face contact people have with 'other groups', the less likely they are to be prejudiced.

Back to boys-are-better-at-maths stereotype. A recent analysis of maths test scores from seven million US students, grades 2 to 11, has found no meaningful difference in average performance among boys and girls. Another study analysed data from 308 studies conducted in 30 countries (though primarily in the US) from 1914 to 2011. The data representing more than one million students from primary school to college proved that girls are just as capable as boys doing maths and science. In fact, female students actually outperformed their male counterparts not only in maths and science but in all subjects. Yet the stereotype persists. The stereotype demoralises young girls who might want to pursue careers in maths and engineering.

Social psychologists call it stereotype threat. This is the idea that by merely bringing to mind a stereotype about ourselves we might accidentally act in a way that confirms the stereotype. An action opposite of the very stereotypical actions we hoped to avoid. A US study of 276 Grade 1 children (average age 6.5 years) showed, for the first time, the power of stereotype threat even in children too young to consciously endorse stereotypes. The children associated boys with maths and girls with language. It's disturbing to see girls as young as six feel the pull of the subconscious stereotype that can push down their maths ability.

What do you do when you witness an actual instance of prejudice? Research shows that you are better off doing something rather than doing nothing. As Charlotte Brontë wrote in *Jane Eyre* prejudices grow 'firm as weeds among stones', those weeds growing firmly among stones in your garden may take over your garden if you do nothing about them. What's better? A dark world overrun by the same foul weeds or a bright world overflowing with a variety of fragrant flowers?

Chapter 18

Would I lie to you?

learn we're all natural-born liars/*unlearn* to make lying a habit

> 'What did she lie about?'
> 'When I say everything, I mean *everything.*' Reiko gave a sarcastic laugh. 'When people tell a lie about something, they have to make up a bunch of lies to go with the first one. *Mythomania* is the word for it. When the usual mythomaniac tells lies, they're usually the innocent kind, and most people notice. But not with that girl. To protect herself, she'd tell hurtful lies without batting an eyelash.'
>
> — Reiko Ishida to Toru Watanabe
> Haruki Murakami, *Norwegian Wood*

If you have not tried it before, now's the time you tried this popular test, devised in 1984 by American psychologist Glen Hass, to find out whether you are a good liar. Extend the index finger of your dominant hand and within five seconds trace a capital Q on your forehead. If you have drawn the Q so that it can be seen by someone facing you with the tail of the Q on the left side of your forehead, you are aware how other people see you and you are a good liar. If you have drawn the Q so that you can see it yourself with the tail on the right side of your forehead, you're an introvert and not a good liar.

Using fMRI scanners neuroscientists have discovered that we use different parts of the brain when we lie and when we tell truth. Neurons in seven areas of the brain fire up when we tell a lie; neurons in only four areas fire up when we tell truth. It seems that the brain has to work harder to tell a lie. Mental conflict arises when we tell a lie and there is increased demand for motor control when suppressing a truth. No wonder neurons in your brain, not your pants, are on fire when you tell a lie.

The stress caused by lying produces some involuntary physiological changes: more sweating, slow breathing and a brief drop in heart rate. These are also the symptoms of fear and anxiety. In 1921 John Larson, a University of California medical student, invented a machine (now known as the polygraph or lie detector) to measure these physiological changes. In a modern polygraph rubber tubes placed around chest and stomach measure respiratory rate, two small metal plates on the finger records sweat and an electrocardiogram (ECG) registers the heart rate. During the test the operator asks a series of true-false questions related to a crime. Advocates of polygraph test say that it's highly accurate. To critics it's as imperfect as measuring Pinocchio's nose. Why this sarcasm? The extremely artificial environment in which a polygraph test is performed makes subjects anxious and they experience sweaty palms and racing pulses.

Can 'innocent anxiety' confuse fMRI? It's too early to say whether fMRI would make a perfect lie detector. At this stage, neither polygraphs nor fMRI can detect a liar with 100 per cent accuracy.

The most complicated form of lying is pathological lying. In a comparative study of self-described pathological liars and volunteers who had no history of deception, researchers have found that pathological liars had 22 per cent more white matter in the prefrontal regions of their brains. Prefrontal regions control decision making and judgment, and white matter makes up the wiring among neurons (which are collectively called grey matter). Pathological lies have information that appears correct and yet is not true. Researchers speculate that it's much easier for such people to lie because the excessive white matter creates an abundance of connections. In other people neural connections are compartmentalised. Perhaps this explains the mind of the mythomanic girl Reiko Ishida is talking about.

We must not confuse pathological lying with confabulation (making up fictitious memories without a conscious effort to deceive or lie). Here are two real-life cases from the files of other doctors. When asked to move his left arm, a stroke patient suffering from the paralysis of the left side of his body, replied with absolute conviction that he had no problem with his arm, despite the clear evidence he was unable to move his arm. Another patient asked about his surgical scar, explained that during the Second World War he surprised a teenage girl who shot him three times in his head, killing him, only for surgery to bring back him to life.

These patients are not knowingly telling lies. In fact, they are engaging in, what Morris Moscovitch, a Canadian neuropsychologist, calls 'honest lying' as they are not aware that their statements are false. They don't know that they don't know what they are claiming.

We all love telling tall tales about ourselves, but some people do really believe in their false stories. To some extent, we all confabulate when we try to rationalise our decisions or justify our options. But clinical confabulation is a type of memory problem.

Most of us are reluctant to admit our failures and mistakes and look for scapegoat. We try to justify our stance even if it means telling lies and more lies. Self-justification is not merely lying to others, it's lying to ourselves. The eminent American social psychologist Elliot Aronson believes that self-justification is more dangerous and more insidious than explicitly lying because 'we are not even aware a mistake was made, let alone that we made it.'

The mind's mechanism behind self-justification is powered by cognitive dissociation: the conflict between two opposing cognitions arouses a contrary psychological stance—called dissonance—which in turn motivates activities designed to reduce dissonance. In other words, it's a state of mind when we oscillate between conflicting ideas, attitudes, beliefs and opinions. According to Aronson, dissonance produces mental discomfort, ranging from minor pangs to deep anguish; people don't rest easy until they find a way to reduce it.

This process of reducing dissonance revs up self-justification. For example, when a teacher exhorts her students to use recyclable shopping bags as plastic bags harm the environment, students expect that the teacher always uses recyclable bags. If a student finds the teacher outside a supermarket with plastic bags full of grocery, points at the bags and smirks, two opposite ideas would cause tension in the teacher's mind: 'I'm a respected teacher and believe in what I teach' and 'I carry my shopping in plastic bags'. Most probably the teacher would ease this momentary mental distress by telling a little lie, 'Oh, left my green bags at home.' If the dissonance is between 'I'm a responsible person and know that drink driving is dangerous' and 'I drive after one too many drinks at the pub' you are forced not to drive from the pub or to find flimsy excuses for driving after drinking. That's how cognitive dissociation drives self-justification.

Self-justification has its uses: we do not spend time worrying and fretting. But the mechanics of self-justification see to it that we become more and more enmeshed in our decision and less and less able to consider the possibility that they are wrong.

Machiavelli, the 16th-century Italian exponent of the art of the politics of duplicity, certainly knew a thing or two about the impression of misinformation on people's memories when he said, 'Throw mud enough and some will stick.' The sticking power of the Machiavellian mud of misinformation has now been enforced by psychologists: lies have a lasting impact on our memory and despite the best efforts to correct wrong facts, they cannot be completely erased. Our brains hold on to lies even after they have been proved wrong; it occurs even if the retraction of lies is understood, believed and remembered. No matter what cock and bull story you tell, some lies will linger on in listeners' memories. It's important to get your story absolutely right in the first place.

Do you think that the following statements are correct? If your answer is yes, you're in the company of people from 75 different countries and 43 different languages who believe in these stereotypes of liars:

Liars avoid eye contact
Liars shift their postures
Liars touch and scratch themselves
Liars are nervous
Speech of liars is flawed

These stereotypes give you little ability to detect lies when someone is lying, but many liars behave quite opposite to what we expect. Still, there is some emotional leakage when we lie. Psychologist Leanne ten Brinke of the University of British

Columbia says that liars' faces are emotionally turbulent, swinging between positive and negative expressions, in marked contrast to the neutrality often displayed by someone telling the truth. Other studies show that eyes don't lie: when you lie, your pupils dilate.

We all tell little lies to enhance our self-image and still are able to retain the self-perception of being an honest person. But little white lies often lead to big black lies. Very soon you may find yourself turning into a mythomaniac. It's only a step away from becoming a pathological liar. The best way to avoid this trap is to learn to tell the truth with tact—or a touch of self-deprecating humour.

Chapter 19

To cheat or not to cheat

learn why we cheat—and when/*unlearn* cheaters have a guilty conscience

> When you tell a lie, you steal someone's right to the truth. When you cheat, you steal the right to fairness.
>
> Khaled Hosseini, *The Kite Runner*

We all know cheating is unethical behaviour; a moral wrongfulness like betrayal, deceiving, disloyalty, disobedience, coercion, exploitation, promise-breaking and stealing. Yet people cheat all the time. Our decision to cheat or not to cheat occurs at a subconscious level. It's a result of subtle behavioural influences of which we are not even aware.

Even time of the day can influence our ethical behaviour. As the day wears on our ability to show self-control to avoid cheating or lying wears down. We are more honest and ethical in the morning than in the afternoon, say researchers Maryam Kouchaki of Harvard University and Isaac Smith of Utah University. They studied the phenomenon, which they have dubbed 'morning morality effect', by giving a series of tests to

327 college-age men and women. In one of the tests, students were shown various patterns of dots on a computer and were asked to identify whether the dots were displayed on the left or right side of the screen. They were paid small amounts of money based on which side they determined had more dots, but the amount was 10 times more for selecting the right over the left. They reported their own scores, giving them an opportunity to cheat. People who participated in the morning session were less likely to cheat than those who took part in the afternoon session.

Another test was designed to test students' moral awareness in the morning as well as in the afternoon. They were asked to fill spaces in word fragments such as '– –RAL' and 'E– – –C– –'. In the morning, students were more likely to fill in the words 'moral' and 'ethical' while in the afternoon they tended to choose 'coral' and 'effects'. Something as mundane as the time of day can lead to a systematic failure of good people to act morally. This effect is linked to our body's energy levels, which inevitably declines as the day goes on. A good reason to make ethical decisions in the morning—and to get up a little earlier.

Thinking about money makes people more likely to cheat, but thinking about time keeps them honest—a conclusion drawn from a series of experiments by Francesca Gino of Harvard Business School and Cassie Mogilner of the University of Pennsylvania. They wondered whether boosting self-reflection might be one way to encourage people to follow their moral compass. In their experiments participants completed various tasks—word scrambles, searching for song lyrics and counting—designed to implicitly activate the concept of money, time or something neutral. Disposing participants to think about money made them cheat more. Thinking about time, on the other hand, prevented them from cheating. The results can be explained by self-reflection: focusing on time leads people to reflect on who they are.

Most of us will cheat when given consciously unnoticeable behaviour nudges. Chen-Bo Zhong of the University of Toronto and colleagues have found that even lighting could affect behaviour. In one of the experiments, participants in a dimly lit room cheated more than those in a well-lit one. In another experiment, some participants wore sunglasses and others clear glasses while interacting with a stranger in a different room. Each participant had $6 to give whatever amount they liked to the stranger. Participants wearing sunglasses behaved more selfishly by giving significantly less than those wearing clear glasses. The researchers say that darkness may induce a psychological feeling of illusory anonymity, just as children playing 'hide and seek' will close their eyes and believe that other cannot see them, the experience of darkness, even one as subtle as wearing a pair of sunglasses, triggers the belief that we are hidden from others' attention and inspections. Try taking your sunglasses off when you approach someone rattling a charity tin in front of you.

Whether you are a believer or nonbeliever the amount of money you are likely to drop in the charity tin—not wearing sunglasses, of course—would be the same. A study by Azim Shariff, a psychologist at the University of Oregon, found no difference in the ethical behaviour of believers or non-believers. But those who believe in a compassionate God are more likely to cheat than those who believed in a wrathful God. The result is in line with other studies that show more revengeful images of God are related to a rigid morality, while people with kind images of God tend to have a relative morality.

Your willingness to cheat may or may not be linked to your view of God, but if you love to stand in 'expansive' postures, either explicitly or inadvertently, you are more likely to accept money you weren't owned. (*See also* 'Sprawling on a big chair can make you dishonest' section in the story 'Let your body do the thinking'.)

A subtle cue may turn you into a cheat, but cheating is a good thing, sometimes. In one study Francesca Gino (mentioned above) and her colleague Scott Wiltermuth asked a group of volunteers to complete a maths puzzle in which multiple columns of figures were added in multiple ways. They were told that they would be paid for each correct answer and they would be scoring themselves. Here was an opportunity to earn free money, if they were inclined to cheat. They were also given another task in which they were presented with three sets of words (for example, sore, shoulder, sweat) and asked to come up with a fourth (cold, in this case). This task, known as a remote association test, measures creativity. Almost 59 per cent of the participants cheated by inflating their score on the maths test. More interestingly, those who cheated did significantly better on the word test. Cheating encourages creativity because breaking rules frees up the mind.

Contrary to popular belief that cheating gives us a guilty conscience, cheating actually triggers a positive effect, a 'cheater's high'. Nicole Ruedy of the University of Washington and her colleagues conducted five studies into how cheaters feel about their actions. They asked participants to take a series of word and number tests and graded their performance. In one of the tests to solve anagrams, participants were told that they would be paid $1 for every one they solved. At the end of the test they were given the answer sheet and told to score their results. They also completed a questionnaire about their mood after the test. More than 40 per cent 'cheated' by making changes to their work sheets after seeing the answers. Interestingly, those who cheated actually felt happier than they had before the experiment, while those who didn't cheat showed no change. In another test which had no financial incentive to cheat, a similar number of participants still cheated, showing it was the psychological excitement rather than money that motivated cheating.

Across all their six experiments, the Gino and Wiltermuth found that people who cheated on different tasks consistently experienced more positive feelings than those who didn't, even though they said that they would feel guilty and have increased levels negative feelings after engaging in unethical behaviour. When people have a chance to cheat even in a relatively minor way and get away with it, the buzz they get from cheating seems to be irresistible.

It's much easier to get away with cheating in the cyberspace. But online infidelity is as damaging to a relationship as real-life infidelity. Research reveals that men and women are just as likely to have cheated both online and in real life while in a serious real-life relationship. In addition, older men are more likely than younger men in real life. People do believe that cheating on a partner is wrong, but they downplay it and minimise its relevance to their sense of self. This makes them feel better about themselves.

'Cheating and lying aren't struggles, they're reasons to break up.' (Patti Callahan Henry, *Between The Tides*). But not so quickly.

Chapter 20

Don't just chalk it up to circumstances

learn to find meaning and purpose in your adverse circumstances/*unlearn* to blame your circumstances for what you are

'People are always blaming their circumstances for what they are. I don't believe in circumstances. The people who get on in this world are the people who get up and look for the circumstances they want, and if they can't find them, make them.'

– Vivie to Mrs Warren
George Bernard Shaw, *Mrs Warren's Profession*

Cleanthes the Water-Carrier. A name that still shines through philosophy books after more than 22 centuries. I have hired him for a day to deliver pails, not filled with water but inspiration and motivation, to all those who are exasperated by the challenges of their circumstances. The first pail he pours over is on a dirty stone in my tiny garden. As the dirt wipes away he exclaims, 'Ah, it's a sundial. Did you know the motto of the sundial?' I shake my head. He continues, 'I record only hours of sunshine.' Make it the motto of your life, he winks.

Cleanthes, a tall muscular man, had spent his youth as a professional boxer in his native Assos, a coastal town in Turkey. In his early 50s he decided that he no longer wanted to grapple with men but with ideas. In around 275 BC he arrived in Athens with only one aim: to sit at the feet of Zeno, the Stoic.

The Classical Age had ended with the death of Alexander the Great in 323 BC and Athens was no longer the city renowned for its architectural grandeur. The city, most of which lay to the northeast of Acropolis, was full of shabby houses. It was totally dry and not well-watered.

With only four drachmas in his meagre bundle of belongings, Cleanthes had to find some work immediately. But he wanted to spend his daylight hours in the academy of Zeno. The only night work he could find was to fill pails of water from a public well and deliver them to houses in the narrow, winding streets. It was work done either by slaves or destitute housewives—in broad daylight. Delivering water in the late hours of night was not welcomed by householders. But he found work in the gardens of rich Athenians; torches were expensive and he had to work in the dark.

The work enabled him to attend the academy of Zeno. He couldn't afford papyrus, so he wrote down Zeno's lectures on oyster shells and ox bones. He continued working as a water-carrier even when he became well-known to Athenians for his simple life but high moral quality. Once, as he was talking to some youths in a public square, the wind blew his cloak and showed his naked body without any shirt. He wasn't ridiculed by his fellow Athenians, but applauded for his frugal and simple life. Like Zeno he wanted to achieve inner peace by being moderate in everything, the Stoic way. After Zeno's death he became the head of the academy and continued for 32 years until his death at the age of 99.

Once someone asked him why he drew water, he replied: 'Is drawing water from the well all I do? Do I not dig? Do I not water the garden? Or undertake any labour for the love of philosophy?' He had made sense of his suffering. He had found meaning in his circumstances. He had given his life the Stoic structure: we can all give our lives coherence in spite of all the defeats and difficulties that might threaten that coherence. His life was now a journey towards understanding Stoicism and practising it even though it might be hard or even impossible to achieve it fully.

Suffering at Auschwitz was of an unimaginably high dimension compared to that of the hard labours of Cleanthes. Viktor Frankl was a Viennese psychiatrist who not only survived Auschwitz but tried to answer the question: 'What makes people survive.' He writes in his gripping book, *Man's Search for Meaning*: 'When we are no longer able to change a situation—just think of an incurable disease such as inoperable cancer—we are challenged to change ourselves.' In the concentration camp, he continues, 'everything can be taken away from a man but one thing: the last of the human freedoms—to choose one's attitude in any given set of circumstances, to choose one's own way.' The ancient Stoics recognised this ultimate freedom. So did Cleanthes, and he exercised it.

In 2006 the Royal Institution of Great Britain named *The Periodic Table*, a profound mixture of science and personal reminiscences, the best science book ever. Its author Primo Levi, an Auschwitz survivor who became one of the greatest writers of the 20th century, also found a purpose in his circumstances: an intense desire to understand his surroundings. In an interview with American novelist Philip Roth in 1986, a year before his death, he recalled: 'I remember having lived my Auschwitz year in a condition of exceptional spiritedness ... I never stopped recording the world and people around me, so much that I still have an unbelievably detailed image of them. I had an intense wish to understand, I was constantly pervaded by a curiosity that somebody afterward did, in fact, deem nothing less than cynical: the curiosity of the naturalist who finds himself transplanted into an environment that is monstrous but new, monstrously new.'

The essence of Frankel and Levi's haunting experiences of the Holocaust is: those who find meaning and purpose in their adverse circumstances survive, and those who fail—do not.

We suffer pain due not to circumstances themselves, but our own illusions, says Bion of Borysthenes, a philosopher and a contemporary of Cleanthes.

Cleanthes, he was no Einstein. He was painfully conscious of his intellectual limitations: 'An old man with grey hairs and no wits,' he once joked about himself. He was nicknamed ass because of his patience and endurance and a certain slowness and dullness of intellect. But he had an abundance of imagination (extant fragments of his verses bear witness) and motivation (his whole life is the testament). Motivation gives purpose and direction to our actions; it provides the will to win. Adverse circumstances do not diminish motivation; they can even have the opposite effect—as the Dalai Lama says 'suffering increases inner strength'.

Say goodbye to the stocky old man and enter *le hangar* on a wintery Paris morning in 1898 to say hello to a slender young woman whose legendary determination to achieve her goal in the face of adverse circumstances has changed our world forever. Many 'd' words would come to your mind to describe *le hanger*—dull, damp, dirty, dingy, drafty—but not despair. Determination—the *determination* of a young woman—is the word that rules *le hangar*, a ramshackle wooden shed in a derelict backyard on rue Lhomond.

Once used as a morgue by students from a nearby medical school to dissect cadavers, the shed has a dirt floor covered with a scattering of asphalt, ruined brick walls, ill-fitting windows and patched glass roof. When it rains, the roof leaks. Drops hitting the floor on worn kitchen tables make soft but nerve-racking noise. A cast-iron stove with a rusty pipe delivers so little heat that in the depths of winter the temperature doesn't rise above six degrees Celsius. That's chilly 43 degrees in Fahrenheit. If this cold doesn't make you shiver, the thought of the ghosts of cadavers haunting the shed would freeze you.

When chemist Wilhelm Oswald travelled from Germany to meet the quiet young woman working in the shed, he thought that he had been shown a stable by mistake. 'If I had not seen the worktable with the chemical apparatus I would have thought it a practical joke,' he recalled later. A blackboard on the wall also reminded him that *le hangar* was indeed a laboratory.

The work the 31-year-old woman, wearing an overcoat to keep her warm, is doing is backbreaking. *La hangar* is filled with large sacks each containing as much as 20 kilograms of pitchblende ore. Each sack is ground, dissolved, filtered, precipitated, collected, re-dissolved, crystallised, recrystallised. As the shed has no chimney to exhaust noxious fumes, this work has to be done in the courtyard outside. In her own words: 'It was exhausting work to carry the containers, to pour off the liquids and to stir for hours at a time, with an iron bar, the boiling material in the cast-iron basin.'

If the balding Bard would have seen the cast-iron basin he would have described it as a witch's cauldron, as he did centuries ago, bubbling with blood of bat, juice of toad, scale of dragon, tooth of wolf, gall of goat, gut of salt-sea shark, root of hemlock dug in the dark and slips of yew tree slivered in the moon's eclipse. It was, in fact, bubbling with hope, hope of discovering something new.

The woman intensely stirring the brew is constantly thinking about her one-year-old daughter who is being looked after by her father-in-law, a retired doctor. The love for her daughter is not diminished by her determination to achieve her goal.

She has only one goal: to find a new element in the mountain of pitchblende ore she is refining. This goal had helped her to find meaning in her exhaustive work in inhospitable surroundings. The mountain of motivation in her mind is higher than the mountain of pitchblende in her courtyard. You know her: Marie Curie, the 'radium woman', the discoverer of radioactivity and the first person to win two Nobel prizes. Her mountain of pitchblende ore eventually yielded not one but two new elements: polonium and radium.

You are now in a warm kitchen of a small house at Woburn, Massachusetts in the winter of 1839. If you feel like having a cup of coffee, it'll be hard to find jars of coffee and sugar as the bench top is crowded with all sorts of stuff: castor oil, ink, magnesia, quicklime, witch hazel; in fact, any cheap material Charles Goodyear, a poverty-stricken American inventor who has just come out of a debtor's prison, could lay his hands on to mix with rubber. Natural rubber gets runny and sticky in the hot weather and turns stiff and brittle in very cold weather. Goodyear's obsession is to find a process to make rubber stable.

One night he spills a mixture of rubber and sulphur on the top of a hot potbelly stove. To his astonishment, the rubber doesn't melt but turns into a gummy mass. 'As I was passing in and out of the room, I casually observed the little piece of gum which he was holding near the fire, and I noticed also that he was unusually animated by some discovery which he had made,' his daughter recalled. 'He nailed the piece of gum outside the kitchen door in the intense cold. In the morning he brought it in, holding it up exultingly. He had found it perfectly flexible as it was when he put it out.'

Goodyear had discovered a process—vulcanisation (after Vulcan, the Roman god of fire), as it later came to be called—that made rubber flexible and usable, regardless of heat or cold. Vulcanisation has made our world rubbery by making tyres, belts, boots, balls and balloons and numerous other rubber products possible.

Though hailed as one of the greatest accidental discoveries, it was the result of the dogged determination of a man who was neither a chemist nor a scientist. Like Newton's falling apple, he wrote in his memoirs published in 1855, the hot stove incident held meaning only for someone 'whose mind was prepared to draw an inference' That meant, he added, the one who had 'applied himself most perseveringly to the subject.'

The story of his perseverance began in 1834 when he recognised raw rubber's valuable properties of elasticity, plasticity, strength and resistance to water but its instability in hot or cold weather made it useless. He endured years of extreme hardship to find a process to make rubber stable. He ran into debts to feed his family and spent time in debtors' prison. Even while in prison he pursued his investigations with passion, asking his wife to bring him a lump of rubber, a rolling pin and some chemicals. With the permission of the sympathetic prison superintendent he turned his prison cell into a laboratory. When he got out of the prison, he continued his experiments, which mainly involved kneading various chemicals into raw rubber. Even after his kitchen discovery it took him five years to perfect the process of vulcanisation, which was patented in 1844. Goodyear Tire Company was founded 38 years after Goodyear's death in 1860 (he left his family nothing but a debt of 200,000 dollars); it's not related in any way with Goodyear or his descendants though it honours his name.

The most admirable thing in Goodyear's life was his clarity of his goal. The worst his circumstances became, the stronger his resolve grew. He had the ability to cope successfully with life's misfortune, tragedies, traumas and setbacks. He could adapt to adversity.

Cleanthes, Curie and Goodyear, they all had set goals they wanted to achieve but were open-minded about the future.

Above all, they also found meaning and purpose in their challenging circumstances.

Now, from scientists to science of purpose-driven life. It has long been known that people who have a higher purpose in life (also known as eudemonic wellbeing) are relatively resistant to disease as well as hardships. Frankel also observed that those with a sense of meaning in life were better able to cope with the horrific circumstances in the concentration camps.

Patricia A. Boyle and her colleagues at Rush University Medical Center in Chicago have followed cognitive health of 900 elderly people for seven years. At the study's outset, the participants were given a questionnaire to gauge their level of sense of purpose in life (defined by researchers as 'the tendency to drive meaning from life's experiences and to possess a sense of intentionality and goal directedness that guides behaviour'). At the end of the study, researchers found that those at the top of 10 per cent on the purpose-driven measure were about 2.5 times more likely *not* to have been diagnosed with Alzheimer's during the seven-year study period, compared with those in the bottom 10 per cent. These more purpose-driven participants also were about 50 per cent more likely to have avoided mild cognitive impairment, an intermediate state of decline that frequently leads to Alzheimer's.

Comfortable circumstances may bring ephemeral happiness but not eternal serenity that comes only when you face the challenges of circumstances by changing yourself.

Chapter 21

Why not the day after tomorrow?

learn Mrs Micawber's poor papa's maxim: procrastination is the thief of time; and the wisdom of *The Time Traveler's Wife* (Audrey Niffenegger): 'Once you've lost time you can never get it back'/ *unlearn* the habit of procrastination is irremediable

> 'My dear Micawber!' urged his wife.
> 'I say,' returned Mr. Micawber, quite forgetting himself, and smiling again, 'the miserable wretch you behold. My advice is, never do tomorrow what you can do today. Procrastination is the thief of time. Collar him!'
> 'My poor papa's maxim,' Mrs. Micawber observed.
>
> Charles Dickens, *David Copperfield*

Your brain is washed in dopamine, a chemical that controls the brain's sense of reward and pleasure. Dopamine neurons are revved up and so is your motivation. But motivation is in the mind; turning motivation into action requires goals that take you beyond your current circumstances. At the same time, once you set a goal and start working towards it your motivation—and life satisfaction—also increases. Achieving a goal requires more than old-fashioned notion of willpower. Willpower may help you control your emotions and impulses but reaching a goal requires an achievable action plan.

Setting a realistic goal requires some effort. We're not talking about your short-term new year's resolutions such as to go on a diet or quit smoking; we're talking about long-term lifestyle changes. First, make sure that the goal is your own; it's not based on someone else's expectations. Think about what motivates you.

The next step is to visualise your goal: visualise the work needed to be done and the specific obstacles you will face; imagine yourself carrying out your plans and what your life will be like once you reach your goal. Visualisation makes the goal seem more tangible and tempting. Studies suggest that imagining both the successful result of your efforts and the specific obstacles you will face—called mental contrasting—help people tackle challenges of circumstances more enthusiastically.

You have your action plan ready and then instead of shouting *carpe diem* (seize the day) you mumble, I'll do it tomorrow. It's the moment all familiar to most of us: the urge to delay the task. Procrastination (from a Latin word meaning 'to put off until tomorrow') is the intentional delay of due tasks. It keeps most of us imprisoned in our present circumstances. Pleasurable social activities have immediate rewards, but benefits of your dreary action plan are distant. How to conquer procrastination?

Well, if I do not feel like writing this book, I can't act like Victor Hugo. The French novelist practised a novel way of dealing with procrastination: he would ask his valet to hide his clothes so that he would be unable to go outside when he was supposed to be writing.

Let me try a different approach, which comes from University of Calgary industrial psychologist Piers Steel (thank you, sir, for letting me to keep my clothes on). The desirability of a task, or utility (U), depends on four factors: the expectation of success (E), the value of completing the activity (V), the delay until reward (D), and the personal sensitivity to delay (I). The equation relating these factors is: U = EV/SD. If I expect to succeed in writing this book (a higher E), its utility (U) will increase. If a publisher promises a lucrative contract, the value of completing the task (V) will also increase U. All I need to do now to keep U high is to make sure that S and D shrink. If I'm impulsive or lack self-control (high S) or the reward lies in the far future (high D), the chances of finishing the task are low. Therefore, to shrink S, I must have high motivation to finish the book; a shorter deadline will help in keeping D low. The printed or e-book you are reading proves that I have not practised 'the art of keeping up with yesterday', as some wit has described procrastination.

In your brain, the limbic system controls automatic functions such as telling you to pull your hand away from a flame. The prefrontal cortex, the 'executive' region of the brain, integrates information and allows you to make decisions. It's not automatic; you must kick it into gear ('I've to finish reading this page of the book'; once you're not consciously engaged in reading the limbic system takes over). Studies suggest that procrastination is a self-regulation failure because it has become automatic and ingrained.

To increase the brain's executive function and decrease procrastination, Timothy A. Pychyl of Carleton University in Canada advises to focus on the problem of 'giving in to feel good' by first developing an awareness of this process and its negative effect on achievement. He also suggests using short goals that build on one another with regular deadlines and feedback. It's easy to procrastinate when goals are large and the path to them long and fuzzy, he says.

Here's another strategy to prevent procrastination: 'implementation intentions', where and how you will strive to achieve your goal. Goal intentions are commitments to act ('I intend to reach Z': study Stoic philosophy, analyse pitchblende ore to find a new element or simply to write an essay). Implementation intentions are dependent on goal intentions and specify the when, where and how of responses leading to goal achievement. Says Peter Gollwitzer of New York University, a foremost authority on implementation intentions: 'They have the structure of "When situation *x* arises, I will perform response *y*!" and this links opportunities with goal-directed responses.' For example, to achieve the goal intention of completing your essay, your completion intention is: 'I'll sit down at my desk tonight at 8 to make an outline of my essay'. Once you have started, avoid getting derailed ('If my mobile rings, I'll ignore it'). You also know when to look at different alternatives to achieve your goal ('If feedback from my tutor is disappointing, I'll change my strategy') and when to stop working too hard ('I'll now watch TV and look at my essay again in the morning').

Most goals are never achieved—unless you make sure they do. But one success erases many failures. Implementation intentions can have an almost magical effect upon you to achieve that one success. The mechanism in the brain is simple: When you sincerely decide 'I'll sit down at my desk tonight at 8 to make an outline of my essay', you hand over the decision from your conscious mind to the unconscious mind. Now the fickle conscious mind sits idyll and the reliable unconscious takes over the control of the executive region of the brain. You will be unconsciously at your desk at 8 pm to plan your essay. In implementation intentions the good old willpower does help.

Mr and Mrs Micawber had strong views on procrastination, but what would they think about pre-crastination? Research by David Rosenbaum of Pennsylvania State University and colleagues hasn't bothered to answer this question but it suggests that people often opt to begin a task as soon as possible just to get it off their plate. Pre-crastination is a term used by them to describe the tendency to hurry up to complete a task as soon as possible. In one of their nine experiments, all which had the same general setup, they asked university students to pick up either of two buckets, one to the left of an alley and one to the right, and to carry the selected bucket to the alley's end. In most trials, one of the buckets was closer to the end point. Contrary to their expectation, students chose the bucket that was closer to the start position and carrying it farther than the other bucket. The researchers say that this seemingly irrational choice reflects a tendency to pre-crastinate. 'Most of us feel stressed about all the things we need to do—we have to-do lists, not on just slips of paper we carry with us or on our iPhones, but also in our heads,' says Rosenbaum. 'Our findings suggest that the desire to relieve the stress of maintaining that information in working memory can cause us to over-exert ourselves physically to take extra tasks.' The researchers says that most of the people they tested pre-crastinated, so they think procrastinating and pre-crastinating might turn out to be two different things.

The time you'll take to turn over the page to start reading the next story would determine whether you like to procrastinate or pre-crastinate. If you dump the book, it's called giving up. You may do so if 'you wanna fly, you got to give up the shit that weighs you down' (Toni Morrison, *Song of Solomon*).

Chapter 22

You may now brag about yourself

learn talking about self is an inherently pleasurable activity/ *unlearn* sharing your thoughts and experiences on social media has no long-term benefits for your wellbeing

> 'Speak,' [Mr Rochester] urged.
> 'What about, sir?'
> 'Whatever you like. I leave both the choice of subject and the manner of treating it entirely to yourself.'
> Accordingly I sat and said nothing: 'If he expects me to talk for the mere sake of talking and showing off, he will find he has addressed himself to the wrong person,' I thought. 'You are dumb, Miss Eyre.'
>
> Charlotte Brontë, *Jane Eyre*

Didn't 'dumb' Miss Eyre know that talking for the mere sake of talking and showing off is what most of us do most of the time? We humans are social animals and spend more time talking about ourselves than other topics. On social media platforms such as Facebook and Twitter talking about our likes and dislikes, thoughts and opinions, interests, work, intimate life and so on takes up 80 per cent of our conversation. Why so many of us are compelled to share our lives with others? Simply because talking about ourselves makes us feel good. It's as satisfying as eating good food, getting money or having sex. Even thinking about yourself enriches you.

The idea why people on average spend 60 per cent of conversations talking about themselves fascinates Harvard University psychology researchers Diana Tamir and Jason Mitchell. Like others they also enjoy talking about themselves, but were keen to find out what drives our penchant for self-disclosure. They devised a series of experiments in which they hooked up 195 participants to an fMRI machine and scanned their brain activity as they discussed their own opinions and personality traits and then those of others.

In the first experiment, the participants alternately disclosed their own opinions or judged the opinions of others. In the second experiment, they alternately disclosed their beliefs about their own personality traits or judged the traits of another person. Because the same participants answered the same questions in relation to both themselves and others, researchers were able to compare the results of brain images. They found that the brain regions associated with reward lit up more brightly when the participants were talking about themselves and less brightly when they were talking about someone else. The same regions of the brain are associated with the sense of reward and satisfaction from food, money and sex. The experiments provide clear evidence that self-disclosure is inherently pleasurable.

In a follow-up experiment, the participants answered questions about their opinions and then either shared them with a friend or relative of their choosing or kept them private. Unlike the first two experiments the participants were clearly told whether their responses would be 'private' or 'shared'. This experiment was designed to find out the importance of audience to one's self-disclosure. The results were similar to the first two experiments but with one important exception: the results were 'magnified' when participants believed their opinions would be shared with someone else. In another experiment, the participants were offered a small cash reward for answering questions about other people such as President Obama and a lower reward for answering questions about themselves such as whether they enjoyed snow skiing or liked mushrooms on a pizza. In most cases, participants turned down the higher amount to talk about themselves. 'Just as monkeys are willing to forgo juice rewards to view dominant groupmates and college students are willing to give up money to view attractive members of the opposite sex, our participants were willing to forgo money to think and talk about themselves,' the researchers say

Put simply, Tamir and Mitchell's experiments show that we love talking about ourselves (self-disclosure) and—in the absence of an audience—thinking about ourselves (introspection). And are even willing to turn down money to enjoy this pleasurable activity. The study also helps to explain why people use social media so often. 'I think it helps to explain why Twitter exists and why Facebook is so popular,' says Tamir, 'because people enjoy sharing information about each other.' Why else would you tweet?

Bragging about ourselves gives us many adaptive advantages: it helps build social bonds and social alliances between people; it elicits feedback from others which builds up our self-knowledge; and expands the amount of know-how anyone can acquire in a lifetime. Above all, self-sharing makes us what we are—social animals; and our wellbeing depends on it.

How much self-disclosure is right? Apply the Goldilocks principle ('Ahhh, this porridge is just right') and disclose just the right amount. While talking about yourself on social media, note that emotions travel much faster than other sentiments. Researchers at Beihang University in China have discovered that joy travels faster than sadness or disgust, but nothing is speedier than rage. Jonah Berger, a University of Pennsylvania marketing professor, has discovered that awe is the one emotion that outpaces rage or anger. 'Awe gets our heart racing and our blood pumping,' Berger says. 'This increases our desire for emotional connection and drives us to share.'

A study on self-disclosure on Facebook has found that people who seek attention and those who have existing or new relationships brag most. Surprisingly, the study found that men and women do not differ when it comes to self-disclosure. Other research shows that Machiavellian men and women who are willing to exploit others without any concern with morality do not reduce the amount of information they shared with others, instead they lied about themselves. An analysis of 30-second samplings of 20,000 conversations has revealed that happiness is correlated with talkative people who went beyond the small talk. Just as self-disclosure instils a sense of intimacy in a relationship, substantive conversations instils a sense of meaning in interactive partners. The essence of these three studies is that you would find happiness when you do not brag too much but have honest and deep conversations. Small talks can lead to meaningful conversations—and happiness.

If self-disclosure is an inherently pleasurable activity, what about self-disclosing to God? A pile of studies has now demonstrated that disclosing a personal secret or writing about a traumatic experience—from telling someone to writing it on a piece of paper—has positive health outcomes. A prayer serves as self-disclosure to God.

American psychologists Patrick Bennett and Marta Elliott designed a study to examine the idea that written prayers about difficult life events serve as self-disclosure to God and therefore should have immediate emotional and physical health benefits. Their study made no claim to unravel or explain the supernatural effects of prayer. The researchers asked 150 participants who identified themselves as religious to write narratives about mundane experiences or about traumatic or stressful life events. As participants moved through four writing sessions, their stories showed increased positive emotions and decreased negative emotion in each of the trauma-writing conditions. The researchers conclude that their close analysis of narratives provide considerable support for the idea that prayer serves a self-disclosure to God and may provide health benefits similar to writing about trauma without God in mind.

Another related matter is that of intercessory prayer or distant prayer. Prayer on behalf of others is widely believed to help recover patients from illness. Sadly, no well-controlled clinical trial so far has supported this belief. But again, no study can answer the question whether we should pray for the recovery of our sick loved ones. Unfortunately, medical science does not follow the precise laws of the universe as described in physics textbooks. We must pray (without telling the patient), but should not expect that by some unknown law of quantum physics our prayers will generate cosmic waves that will improve the physical wellbeing of the patient. At least, prayer will make you feel happy, much so if it's peppered with self-disclosure.

Science now approves of talking to yourself, but it has not disproved the intrinsic rewards of listening to others. Learn the art of listening from dogs, as Jerome K. Jerome writes in *Idle Thoughts of an Idle Fellow*: 'They never talk about themselves but listen to you while you talk about yourself, and keep up an appearance of being interested in the conversation.'

Chapter 23

Out of the proverbial box

learn everyone can think clearly, creatively and imaginatively/ *unlearn* to say 'I'm not creative or the creative type or don't have time for the creative stuff'

'... If I were to speak your kind of language, I would say that man's only moral commandment is: thou shall think ... all work is creative work if done by a thinking mind, and no work is creative if done by a blank who repeats in uncritical stupor a routine he has learned from others ...' – from John Galt's very long speech, which provides an introduction to the novelist's philosophy which she calls 'objectivism'

Ayn Rand, *Atlas Shrugged*

The story goes that John Dalton gave his mother a pair of stockings as a birthday present. 'You have bought me a very fine pair of stockings, but what made you fancy such a bright colour?' she asked. 'I can never show myself at meetings in them. They're as red as cherry, John.' Much distressed by his mother's remark, he asked his elder brother who, like Dalton, thought the stockings were of a drab dark-bluish colour. When their neighbours also remarked 'very fine stuff, but uncommon scarlety', Dalton realised that he and his brother suffered from some genetic defect.

In 1798 Dalton, the English chemist who has an honoured place in the history of science, made the first scientific study of colour blindness, which is caused by a deficiency in photoreceptors, a fact Dalton didn't discover (he thought the insides of his eyeballs were tinted blue). A humble and frugal man, he was highly admired by his contemporaries. When he died, the City of Manchester decided to give its famous citizen a state funeral; his body was viewed by more than 40,000 persons and the funeral procession was made up of more than 100 horse carriages and countless persons on foot.

How did Dalton arrive at his idea of the cause of colour blindness? Like all creative people he primed his brain for creativity by wandering the world with an open mind. We can imagine that like most creative people he looked at the coincidence of two persons seeing bright red stockings as blue. We tend to ignore coincidences because we think they are meaningless. Dalton probably tried to correlate these two coincidences and other incidences of colour blindness he may have observed during his search for a reason for colour blindness.

His next step was imagination. When solving a problem, as Einstein has said, 'imagination is more important than knowledge'. Dalton may have used his imagination to answer questions such as: Why did his eyes perceive the colours differently? Did some structure in the eye caused this inability to see colour red as red?

Dalton was intrinsically motivated, meaning he was motivated by the desire to perform the task for the enjoyment it provided, not for money, prestige or power. Intrinsic motivation boosts creativity. Solving a problem also requires resilience: you fail repeatedly until you find the right answer. His motivation and resilience paid for when he experienced the 'Aha!' moment: the insides of his eyeballs were blue, some kind of genetic defect, which made him colour blind. This idea was later proved wrong, but it was a creative solution according to the scientific knowledge of the time.

When people have a eureka moment they are not consciously monitoring what they are thinking. The feeling of finding a solution to a problem is an emotional experience, which can be likened to relief. Studies show that people who unwittingly solve problems show more alpha waves in their brain at the time they find the solution than those who are knowingly thinking very hard to come up with an answer. Alpha waves are associated with relaxed state of mind.

Dalton would have been elated by his insight into the cause of colour-blindness. That eureka moment, like all eureka moments, came with a lot of conviction. An insight may lead to wrong solution if the idea is not carefully evaluated. As mentioned earlier, creativity requires resilience:

most creative people go through many failures before finding a successful solution.

Dalton obviously had the capacity for constructive use of solitude. Research shows that when groups of people interact for the purpose of brainstorming, they significantly overestimate their productivity and produce fewer unique ideas than normal groups of people who generate ideas alone. If you want to be creative, be willing to work alone. Working alone would allow your mind to daydream. Nevertheless, we must not forget that all creativity takes place in a social context. In addition to personal identity, all creative people also need social identity that enables them to continue their creative work in challenging circumstances.

We now have enough experimental proof to say that daydreaming certainly boosts creativity. It's an important tool for creativity. The intense focus on a problem has its advantages, but the relaxed style of thinking leads us to contemplate ideas that sometime seem silly or far-fetched. Such imaginative thoughts might not be practical, but they often are the perfect springboard to creative insights.

You don't have to be a poet, artist or scientist to be creative. You can express your creativity in everyday life: devising a new recipe, landscaping your garden, painting your house, making handicrafts ... the list is endless. Any project you undertake requires thinking, thinking how you will do it.

Creativity is not a gift from the muses only granted to certain people, says Robert Epstein, a renowned expert on creativity. Everyone can enjoy thrills of 'little c' creativity that involves solving everyday problems.

Leave 'big C' creativity, that is about big ideas that change the world, to the likes of Michelangelo, Mozart and Einstein.

In brief, to be creative:

- Know your stuff. Immerse yourself in the problem. Ask questions.
- Keep trying. You are unlikely to solve a big problem at the first try, or the second, or the third ...
- Be ready to imagine the impossible. Many breakthroughs at first seem completely crazy.
- Relax: laugh, daydream, doze.

More ways to catch creative ideas:

Unstick it. To us bicycle is one entity with only one use, but if we think in an innovative way it's a collection of parts and each part can be used in one or more ways. Psychologists call our tendency to fixate on the common use of an object or its parts 'functional fixedness'. This rigid thinking hinders people from solving problems.

American psychologist Tony McCaffrey has developed a systematic way to overcome this obstacle. Suppose you are given two steel rings, a candle and a match to make a figure-8 out of the two rings. Melted wax is sticky but it's not strong enough to hold two steel rings. McCaffrey suggests breaking down the items at hand into their basic parts and then name each part in a way that does not imply meaning. The wick in the candle implies 'wicks are set afire to give light'. But if we think of wick as a string it opens up a new possibility: remove the wick and tie the two rings together.

McCaffrey calls this way of thinking 'generic parts technique' (GPT). In one of his experiments participants trained in GPT solved eight problems 67 per cent more often than those weren't trained. He explains how GPT works: For each object in your problem, break it into parts and asks two questions: (1) Can it be broken down further? (2) Does my description of the part imply a use? If the answer to the second question is 'yes', decouple the name of the part from its use. When you rename it, you can reuse it.

Move your eyes. Contrary to popular belief, right hemisphere is not the one that thinks outside the box. Creativity is much more complex than the right-left brain distinction. In fact, creativity does not involve a single brain region or single side of the brain. It involves many cognitive processes, emotions and neural pathways and we still do not know fully how the creative mind works. Try this simple exercise: move your eyes to follow a target as it moves horizontally left to right for 30 seconds. This exercise is thought to increase interaction between right and left hemispheres of the brain. This collaborative effort between the left and right hemispheres is believed to boost creativity.

Become a collector of coins, stamps, wines, minerals, butterflies or whatever you like. Collectors observe acutely, make fine distinctions and recognise patterns in their collections—mental tools that are necessary for creativity. Recognising pattern is important as it also tells collectors about missing pieces in their collection. A new discovery is usually putting pieces of a puzzle together to find something new.

To kick-start your mind's engine of creativity. Be happy. Happiness prevents laser-like focus, which is not conducive to creativity. Happiness is like seeing the world through the narrow end of a funnel: what you see is brightness.

Or: Travelling to faraway places, thinking about faraway places, communicating with people who are dissimilar or thinking how things might have happened, they all give you some psychological distance, which helps you to transcend the immediate moment in your mind. The further you move away from your own perspective, the wider the picture you are able to consider. This, in turn, makes you more creative.

Or: Move out of a quiet library to a cafe with a bit of background noise (not TV or radio at full blast) can enhance creativity.

Here's an example of how a simple act of 'small c' creativity can energise you with 'big I' inspiration:

There was something so significant about being able to make a gorgeous item of clothing from almost raw materials. It gave her a feeling of her own power, to make something practical and beautiful just by using her own skill and creativity. It inspired her.

Kate Jacobs, *The Friday Night Knitting Club*

Chapter 24

Weighing up your choices

learn to make rational decisions/*unlearn* to make irrational decisions

> 'Thinking. A process by which I use my brain to make a rational decision.'
>
> Becca Fitzpatrick, *Hush, Hush*

When in 1984 Australian medical scientists Robin Warren and Barry Marshall proved that the stomach ulcer isn't a painful chronic condition but an infectious disease that can be cured easily, their idea was ridiculed by critics who said that bacteria can't survive in the high acidic environment of the stomach. To prove them wrong, Marshall swallowed a small dose of the live bacteria. He duly developed a stomach ulcer, which was promptly healed with antibiotics. Warren and Marshal were awarded the Nobel Prize in Physiology or Medicine in 2005, not for this amazing act of self-experimentation but for their amazing discovery.

We do not know what was going through Marshall's mind when he made the decision to swallow live bacteria, but we do know how the mind arrives at a decision by weighing up risks and rewards associated with it. Most of us are pretty hopeless at weighing up risky choices, but definitely not Marshall. Daniel Kahneman, Nobel Prize-winning cognitive psychologist and author of the highly fascinating book, *Thinking, Fast and Slow*, describes two types of decision-making systems in the mind. He calls them systems 1 and 2 for the ease of explaining his ideas but stresses that they are not systems in the standard sense of entities with interacting aspects or parts. Also, they do not refer to particular parts of the brain.

System 1 is automatic, based on intuition it uses limited amount of information and arrives at a decision quickly and effortlessly. Very sensitive to context and emotions, it's the system that makes the instant decision whether to go through a yellow traffic light. If we didn't have this system we would be caught in an endless cycle of analysis, drawing infinite pros-and-cons lists in our heads. Decisions informed by intuition are prone to all sorts of cognitive biases, but they are likely to produce happier outcomes. Intuitive decision-making has also played a crucial role in some of the most remarkable scientific discoveries, including that of Warren and Marshall. Though gut responses help us to concentrate on the right decision, it's best to avoid making important decisions when you are emotional.

System 2 is rule-based and is based on reflective and deep thinking. In contrast to system 1, it's slow, requires effort and conscious thinking. This is the system you engage when you are solving a problem. You don't want to follow your instincts when choosing a mortgage plan. You need to take all pros-and-cons into account to avoid a costly mistake.

Systems 1 and 2 are complementary, so are the two neural systems that are involved in making decisions. When Marshall was thinking about whether his idea of swallowing live bacteria was reasonable or risky, the brain's emotional regions would have become highly active. But system 1 didn't make an instant decision as system 2 was also at full alert. Before Marshall decided to swallow live bacteria the activity in the prefrontal cortex in his brain would have become active. This region is responsible for making decisions. Undoubtedly, the final decision was a well-informed and balanced decision.

You would think that having more options to choose from for your mortgage would make you happy, but too many choices, too much information, don't help you to a make a rational decision. Information overload caused by mind-boggling choices results in information fatigue. Your brain's emotional regions go haywire.

American psychologist Barry Schwartz calls it 'the tyranny of choice'. He says that feelings of happiness initially rise as choice increases but level off quickly. Net result is that, at some point, added choice only decreases happiness. He suggests to settle for a choice that meets your core requirement rather than searching for the elusive 'best', and then to focus on the positive aspects of your choice. The old cliché 'don't expect too much, and you won't be disappointed', he says, is sensible if you want to be more satisfied with life.

You have made your decision, say, about your mortgage. But no decision is foolproof: there's always a risk of making a wrong choice. Whether you make a wrong decision or a right one your brain makes the decision 10 seconds before you realise that you have made it. Researchers have arrived at this conclusion after looking at the brain activity of volunteers while they performed a decision-making task.

You can sharpen you decision-making skills by taking one simple step: wait a fraction of a second before making your decision, say, to eat a doughnut or hit the accelerator or brake when approaching a yellow traffic light. Psychiatrist Tobias Teichert of the University of Pittsburgh and colleagues asked volunteers to make simple decisions very quickly. While looking at sets of dots scrolling across a computer screen, they were told which set was their target and then asked to indicate which direction the target dots were moving when new sets of dots were introduced to distract them from their task. The average reaction times of participants varied between 500 and 600 milliseconds. When the participants had a little more time, even as few as 50 milliseconds to make their decision, their accuracy improved. The team found that it takes at least 120 milliseconds—the time it takes to double-click a mouse—for the brain to switch its attention from task to task. This may not seem like very much, says

Teichert, but this waiting enables us to focus on the most relevant information.

Still worried about a decision you made after waiting a while? Wash your hands and you will wash away your doubt. Washing your hands of something is an old idiom, but scientists say this metaphor is indeed true. In an experiment in which participants chose between things like a pair of CDs or two flavours of jam, those who washed their hands right after making their choice didn't regret their decision.

If you didn't wash your hands after buying this book, do not rush to wash them now. Just be thankful and you won't regret your decision. Psychologist David DeSteno from Northeastern University in the US and colleagues randomly assigned volunteers to one of the three conditions in which they wrote about an event from the past that made them feel either grateful, happy or neutral. The volunteers then chose between receiving an amount of money now or a larger amount after a few months. Volunteers feeling neutral and happy showed a strong preference for immediate payouts (on average, $55 to forgo $85 in three months). Grateful volunteers showed more patience: they required $63 immediately to forgo receiving $85 after three months). The results show that feeling grateful improves people's ability to take the long-term view when making financial decisions. DeSteno rejects the idea that emotions are to be dampened to avoid irrational impulses for immediate gains. 'Emotions exist to serve adaptive purposes, so the idea that emotions would always be a hindrance to long-term success makes little sense,' he says.

It doesn't matter what decision you make, being able to make a decision is a reward in itself. The decision triggers neurons in your brain's pleasure centre to fire up with joy. Just relax and relish the deliberation. Still unhappy? Follow ancient Persians.

The ancient Greek historian Herodotus describes a custom of Persians on making decisions in his book, *The History*. He writes that men would make a decision three times: once while sober, once while drunk and then again when they sobered up next day. If the decision still pleased them, they'd act on it; if not they'd give it up. Conversely, whatever decision they made when sober, they reconsidered it when drunk.

Should we drink to it?

Chapter 25

You foul mouth #$&%

learn swearing has its benefits/*unlearn* bad language can never be good for you

> Some producer actually told Franny that profanity revealed a poor vocabulary and a lack of imagination. And Frank and Lilly and Father and I all loved to shout at Franny, then, and ask her what she had said to that. 'What an anal crock of shit, you dumb asshole!' she'd told the producer. 'Up yours—and in your ear, too!'
>
> – John Berry, the novel's protagonist
>
> John Irving, *The Hotel New Hampshire*

It's not crap. Multilinguals tend to swear in their native language rather than languages acquired later on. Though swearing makes you feel emotionally better in you native language, once you are competent in a language, skill in swearing in that language comes easily. Swearwords have long held a unique and colourful place in all languages.

Swearing is always bad for you whatever language you use. Hell no. Swearing, or cursing, is a form of language that allows us to vent or express strong emotions, positive or negative, such as anger, frustration, surprise or joy. Because of its strong expressive power swearing provides a sense of stress relief. It helps in pacifying our anger and thus becomes a substitute for physical aggression. It also promotes group bonding and helps in eliciting humour. 'It's like a horn in your car—you can do a lot of things with it. It's built into you.' These revealing words come from Timothy Jay, an American psychologist who is definitely not a potty mouth but an expert in the science of swearing. For more than three decades he has recorded 10,000 children and adults swearing spontaneously in public.

Swearing packs an emotional punch because it recruits our emotional facilities to the fullest, says Steven Pinker, a renowned linguistic explorer. 'I think the reason swearing is both so offensive and so attractive that it is a way to push people's emotional buttons, especially their negative emotional buttons ... it always evokes an associated meaning and emotion in the brain. So I think that words give us a little probe into other people's brain.'

Words do tell us a lot about the brain. In 1866, at 45 years of age, French poet Charles Baudelaire suffered a stroke that damaged the left hemisphere of his brain, which literally left him speechless. The only word he was capable of speaking was a now-outmoded French expletive 'Cré nom' (meaning something like 'goddamn' in English). The expletive annoyed the nuns who were looking after him so much that they called in a priest to perform an exorcism. Foul-mouthed but otherwise speechless stroke patients also show damage to the left hemisphere.

This observation has led neurologists to speculate that the language rich in vocabulary, grammar and syntax resides in the left hemisphere, while swear words, prayers, songs and lyrics are stored in the left brain. Swearing is one of a small set of speech functions—'automatic speech'—which is selectively preserved in patients with a damaged left hemisphere. In other neurological disorders when patients experience frustration, they also express their emotions through swearing. People with Tourette syndrome—caused when some neural circuits go awry—make involuntary movements and sounds; some 10 to 30 percent of patients also display the uncontrollable urge to blurt out swearwords. Researchers have even documented the case of a deaf man with Tourette whose swearing was actually expressed in sign language.

Why do most of us use bad language when we stub our toe and are left in pain? Perhaps the question should be framed more appropriately, Why the @&#$ do we swear when in pain? To answer this question a team of neuroscientists from UK's Keele University lead by Richard Stephens asked a group of college students to hold their hand in ice water for as long as they could tolerate pain, to a maximum of five minutes. Students could hold their hands much longer when they were allowed to swear compared to when they did not swear. Their favourite swear words were fu*k and sh*t.

How swearing achieves its pain soothing effect is unclear but it's probably linked to amygdala, the brain's emotion hub. Swearing activates the almond-shaped amygdala, which triggers releases of pain-killing endorphins. The body responds to perceived threat or danger by releasing this 'fight or flight' hormone. Swearing can help us better tolerate pain, says Stephens, but too much swearing in everyday situations can reduce its effectiveness. 'I would advise people, if they hurt themselves, to swear,' he adds. 'But if you really want to benefit from swearing save it up for when it really matters.' Like when a shower unexpectedly starts sprinkling hellish cold water over your head instead of refreshing warm water.

Besides pain relief, Neel Burton, an American psychologist and author of many popular psychology books, also advocates other benefits of swearing: (1) By swearing we show, if only to ourselves, that we are not passive victims but empowered to react and fight. (2) Swearing can be a way of showing that we really mean something or that it is really important to us. (3) Swearing can serve to show that we belong in a certain group and are wholly comfortable with members of that group.

Good or bad, most parents would be concerned about their children swearing. But they might be fighting a losing battle. 'As soon as kids start talking, they pick up this kind of language,' says Jay. 'They're little language vacuum cleaners, so they repeat what they hear.'

Swearing is a lazy language and that's why children find it

easy to pick up. When we use swearwords, we might express a lot of anger of anger or joy but the actual words we say do not mean much. We do not yet know much about what children know about the meaning of words they use. We also do not know whether swearing itself is harmful. When we hear children swearing we assume that they lack discipline or their parents have a relaxed attitude about language.

In today's informal society teenagers find the fuss about swearing unnecessary. F-word is a part of their everyday vocabulary though other swearwords are less common. According to Jean Aitchison, a British linguist, even 'some older people have started to swear in order to seem friendly'.

Friendly or not friendly, simply saying 'hello' says much more about you. New research by University of Glasgow psychologists indicates that we begin to form first impressions based on the tone of voice, specifically one of the quickest and shortest sociable words, 'hello'. The researchers, led by Phil McAleer, recorded 64 students reading in neutral tone an unfamiliar passage, which included a telephone conversation. They then extracted the word 'hello' from each recording and asked 320 different students to listen to the word and rank the voice according to 10 personality traits.

The researchers found that men with lower pitched voices were rated as more dominant. But the opposite was true for women: Those associated with higher average pitch were perceived as more dominant. On the trustworthy scale, men who raised the tone of their voices and women who alternated the pitch of their voices were rated higher. 'It's amazing that from such short bursts of speech you can get such a definite impression of a person,' says McAleer. 'And that, irrespective of whether it is accurate, your impression is the same as what the other listeners get.

In moments of stress, pray or swear? You may pray, but don't swear at Mark Twain, who is believed to have said that under certain circumstances, profanity provides a relief denied even to prayer. Why not just say 'hello'?

Chapter 26

Talking to yourself, again

learn talking to yourself can actually benefit your mental abilities/*unlearn* inner speech—the monologue you 'hear' inside your head—is pointless

> 'I rolled my eyes. 'He's talking to himself. My vote is he's crazy.' He thought about this. 'Maybe he's normal and we're the crazy ones. Maybe everyone should talk to themselves. Maybe we're all just afraid of what we'd say.'
>
> Katie Kacvinsky, *Awaken*

You're definitely not crazy if you talk to yourself. We all talk to ourselves, silently and loudly. As we grow older we talk to ourselves out loud much less often but still sometimes we speak to ourselves audibly. What's the purpose of this nonsensical self-directed speech? Inner speech, inner voice, internal monologue, verbal thought or whatever you call 'talking to ourselves in our head' shapes our thoughts and cognition.

It's difficult to read the content of people's minds but American psychologist Russell Hurlburt has pioneered the ways to investigate our inner thoughts and feelings. He asked volunteers to carry a beeper that made a random beep. Whenever they heard a beep the volunteers had to stop and write down whatever they were experiencing in their minds immediately before the beep. Later, he interviewed the volunteers to learn more about the mental experiences they had recorded. These studies have revealed that about one quarter of volunteers' mental activity appeared to be speaking silently rather than experiencing other visual or emotional phenomena.

When you get up in the morning and say 'what a beautiful morning'. You're talking to yourself. When a friend asks you for advice, you talk to yourself before you talk to your friend. That's also inner speech. The ceaseless inner banter is a component of our conscious self-awareness. 'What we call the self is a story that we continuously write and rewrite in our minds—a narrative that appears to rely, at least in part, on verbal thought specifically,' writes science editor Ferris Jabr in *Scientific American*. 'Merely reacting to the world around oneself by talking to oneself about it may be essential to maintain a cohesive identity that persists through the past, present and future.' Not only does it help to maintain our identity, inner speech helps us to solve problems, plan for the future and learn from past mistakes.

Inner speech is basically thinking in language; it's an exact reflection of our personal voice with our own vocabulary, favourite phrases and idioms. The Russian neuropsychologist Alexander Luria pointed out in the 1970s that language enabled humans to use symbols to represent events and physical objects. Through a process of inner speech humans were freed from the confines of the present to reflect upon the past and plan for the future.

Language is more than just a tool for communicating with others. Stanford University cognitive psychologist Lera Boroditsky believes that the languages we speak affect our perceptions of the world. She compares two incidences to support her assertion. During one of her trips to Cape York in northern Australia she asked a five-year-old Aboriginal girl to point north, she pointed precisely and without hesitation. Later, back in a lecture theatre at Stanford, she asked her audience of distinguished scholars to close their eyes (so they couldn't cheat) and point north. Some of them have come to that very room to hear lectures for more than 40 years, she points out. Still, many refused, most failed to point correctly. Why can a five-year-old in one culture do something with ease that eminent scholars in other cultures struggle with it?

Boroditsky believes that the difference in their cognitive abilities is due to difference in their languages: each language has its own cognitive tool and, therefore, different languages impart different cognitive skills. The way we think influences the way we speak, she says, but the influence also goes the other way. Other research findings also support her view that language is part and parcel of many aspects of thought and, therefore, inner speech.

Research by American cognitive psychologists Gary Lupyan and Daniel Swingley also shows that language is more than a communication tool as it can also help increase perception and thinking. 'I'll often mutter to myself when searching for something in the refrigerator or supermarket shelves,' says Lupyan. So, the psychologists conducted a series of experiments to discover whether talking to oneself can help when searching for particular objects. In one of the experiments, participants were shown 60 colour pictures of objects such as banana, giraffe and guitar and asked to find a particular object. Half of the participants were asked to repeat out loud what they were looking for while the other half kept their lips sealed. Those who talked to themselves found the object slightly faster than those who didn't.

What happens in the brain when we hear the inner speech in the absence of actual sound? A study by Canadian linguist Mark Scott and colleagues concentrated on corollary discharge. It's a signal generated by the brain that prevents confusion between self-caused and externally caused sensations. It explains why other people can tickle us but we can't tickle ourselves: the signal predicts our own movements and effectively cancels out the sensation of tickle. Similarly, corollary discharge for speech is an internal prediction of the sound of our own voice. Scott says that sound we hear inside our heads when we talk to ourselves is actually the internal prediction of the sound of our own voice.

Talking to yourself is more than a kind of internal cheering system—'I can do it', 'I'm almost there'—but even this helps in boosting confidence. A review of more than two dozen studies has found that 'instructional self-talk' has many more advantages: it enhances our attention and helps us focus on the task; it helps us make decisions about what to do, how to do and when; and it controls out emotional and cognitive reactions encouraging us to stay on the task.

What to say when you talk to yourself? 'Tell yourself often enough that you'll fail and you will almost certainly will,' advises American clinical psychologist Clifford Lazarus. 'Tell yourself often enough that you'll succeed and you greatly improve your chances of fulfilment and satisfaction.' Never say 'I can never handle such a complex task' but say 'If I go slowly, step-by-step I'll be fine'.

In Thomas Pynchon's, *Gravity's Rainbow*, Dr Ned Pointsman says, 'Talking to myself, here. Little—sort of—eccentricity, heh, heh.' Talking to yourself may be a sort of eccentricity but it helps us in numerous ways. Yet too much of this wrong kind of self-talk can backfire. Obsessive self-talk about painful experiences carries risk of depression. Just take it easy.

American humourist Franklin P. Jones has said: 'One advantage to talking to yourself is that you know at least somebody's listening.' If you have got a good listener, it doesn't mean that you bore yourself with endless anxious self-talk.

Chapter 27

It's time to catch some zzz...

learn sleep is something everyone does every day but still it's enlightening to know more about sleep/*unlearn* sleep myths such as 'when we sleep, the brain shuts down'

> *Bien haya el que inventó el sueño* (Blessings on him who invented sleep)
>
> – Sancho Panza
> Miguel de Cervantes, *Don Quixote*

Sancho Panza continues to sing praises of sleep as 'the cloak that covers our thoughts, the food that cures all hunger, the water that quenches all thrust, the fire than warms the cold, the cold that cools the heart ... the balancing weight that levels the shepherd with the king, and the simple with the wise.' No wonder the simple and the wise all want to crash into sleep; and while awake contemplate the mysteries of sleep.

Simple Sancho Panza thought sleep 'has only one fault, that it is like death; for between a sleeping man and a dead man there is very little difference'. Wise scientists know that, unlike the dead brain, neurons in the sleeping brain are hard at work: they fire nearly as often as they do in the waking state, and they also consume almost as much energy. During sleep new synapses—the junctions between communicating neurons—are formed and the old ones are pruned. This task helps the brain to clean out obsolete information and consolidate memories.

Strengthening the links between neurons is not the only function of sleep. The brain uses up to 20 per cent of our body's energy, but it's outside the reach of the lymphatic system. Whenever waste is formed in our body the lymphatic system sweeps it up clean. New research shows that the brain has its own cleaning system, which sweeps out the neural trash. Scientists are still a long way away from completely understanding this system, but once they know how it works it would enable them a new understanding of Alzheimer's and other forms of dementia.

But something is obvious right now: if we do not get enough sleep, what happens to trash that piles up in the brain? Are we compromising our mental abilities, if we do not get enough sleep? And why do we need to sleep? Is it just because we get sleepy or are there other reasons for it? Sancho Panza was not worried about such questions about sleep ('all I know is that so long as I am asleep I have neither fear nor hope, trouble nor glory'); we are. Here're some of the things we do know about sleep.

Sleep on it

Our memories need sleep. Most scientists agree that sleep is important for forming and retaining memories. Whenever we hear, see, read or feel something millions of neurons fire in a highly coordinated way. The memory of that event is a group of connections among all those neurons. Whenever any of those neurons is reactivated, the whole group is reactivated to create the memory of that event.

Until recently it was believed that recently formed memories were replayed during sleep, which strengthened neural connections and in the process memories became more sharply etched on the brain. New evidence shows that neurons in the hippocampus, the centre of learning and memory in the brain, fire backwards during sleep. This backward firing weakens neural connections freeing up space in the brain to store new memories on waking. If you want to memorise some new information, the best time to learn it just before going to sleep. The old adage 'sleep on it' is still worth sleeping on.

Learn while you sleep. No, you can't place a pile of dockets from your shopping spree under the pillow and know the total amount of money you have blown away in the morning. But experiments by Israeli researcher Anat Arzi prove that we can learn entirely new information while we sleep. She and her colleagues used a simple form of learning called classical conditioning to teach 55 participants to link smells with sounds as they slept. They exposed sleeping participants to various pleasant and unpleasant smells such as deodorant and rotting fish and played a specific sound to accompany each smell. The sleep conditioning persisted even after the participants woke up: a relevant sound caused them to sniff even if there was no smell. The participants didn't know that they had learned the link between smells and sound during their sleep. Arzi says that sleep learning could lead sleep therapies to modify behaviour to overcome phobias. Other researchers have already used certain smells to lower levels of fear in people by triggering and re-channelling frightening memories into harmless ones during sleep.

Participants in an experiment at Northwestern University in the US learned how to play two musical tunes in time with moving sequence of circles that showed which key to strike. They then took a 90-minute nap, during which they were repeatedly exposed to one of the two tunes during slow-wave sleep, a stage of sleep thought to be important for cementing memories. Tests after the nap revealed that participants performed the tune played during slow-wave sleep more accurately than the one that they were not exposed to. 'If you were learning how to speak a new foreign language during the day, for example, and then tried to reactivate those memories during sleep, perhaps you might enhance your learning,' says co-researcher Paul Reber.

Sleep helps in making better decisions. Taking a night to sleep when you are facing an important decision is the common wisdom, which has now been confirmed by science. During REM sleep, a stage of sleep that is accompanied by rapid eye movements and accounts for 20 to 25 per cent of our sleep, the brain is highly active and creative. This sleep benefit in making decisions is due to changes in underlying emotional or cognitive processes during REM sleep.

Sleep stimulates lateral thinking. Jan Born of the University of Lübeck in Germany and colleagues gave a group of volunteers a series of numbers with a simple rule with which to generate a second series of numbers from the first, and asked them to deduce the final digit in this sequence. The final digit could be calculated immediately by a shortcut, but volunteers were not told about this hidden shortcut. Those who tackled the problem in the evening and returned refreshed after a good night's sleep were more than twice as likely to spot the shortcut as those who had stayed awake. Another group that tried the problem first in the morning and then spent a normal eight hours of the day awake, was just as bad at spotting the trick as the group that stayed awake at night. This ruled out that poor performance wasn't simply due to being tired. It was most likely that the volunteers who found the solution in their sleep were dreaming about the puzzle during their REM sleep, which is associated with dreams.

An example of one of the greatest breakthroughs in science that came during sleep is that of the periodic table. In 1869 Russian chemist Dmitri Mendeleev was struggling with the problem of the order in which to introduce the 61 elements then known in his new textbook of chemistry. He listed the names and properties of the elements on individual cards and began a lengthy game of solitaire (patience), trying to arrange the cards in different ways. Tired, he fell asleep at his desk and dreamed. 'I saw in a dream a table,' he wrote later, 'where all the elements fell into place as required.' The table he saw in his dream eventually became the iconic periodic table.

It's a misunderstanding that the sleeping brain isn't doing anything; it's busy organising memories and picking out the most important information making you to come up with new ideas.

Stay awake for it

Lack of sleep may make us unethical. Lack of sleep may promote unethical behaviour by diminishing self-control. This is the conclusion of a field study that examined unethical behaviour in a variety of work settings. Low levels of sleep and low perceived quality of sleep were both positively related to unethical behaviour, say American management expert Christopher Barnes and colleagues. They believe organisations may benefit from providing work schedules that provide better sleep. Other research has highlighted the impact of poor sleep on workers' health, morale and safety. Brain scans of sleep deprived but otherwise healthy people show that increased activity in the amygdala, the brain's emotional centre, but decreased activity in the prefrontal cortex, the decision-making region of the brain that makes us rational.

There is, after all, something called 'beauty sleep'. Sleep-deprived people may or may not become unethical but they are likely to be perceived as less healthy and less attractive compared with when they are well rested. A study by six Swedish neuroscientists led by John Axelsson involved 23 healthy, sleep-deprived adults (age 18–31) who were photographed and 65 untrained observers (age 18–61) who randomly rated the photographs. The neuroscientists, who apparently compromised their own 'beauty and health' when they lost sleep over their rigorous study, proudly claim in the prestigious medical journal *BMJ* that 'sleep-deprived people were rated as less healthy, less attractive and more tired'.

Is there an underlying message in the above two studies for managers to provide sleep pods along with desks or work benches if they want an ethical and beautiful workforce?

Take points off your IQ. Lack of sleep may or may not dampen your beauty or ethics; it is likely to dampen your children's IQ. A few years ago a team of researcher at Tel Aviv University led by Avi Sadeh sent 75 primary school students home with randomly drawn instructions to either go to bed earlier or stay up late for three nights. Each child was given an actigraph, a wristwatch-like device that recorded how much sleep they were getting. The researchers found that the first group managed to get 30 minutes more sleep a night; the other group got 31 minutes less. When the researchers tested children's IQ after the third day, they discovered that slightly sleepy grade 5 children performed in class like grade 3 children. A loss of one hour of sleep was equivalent to two years of cognitive development. These findings are consistent with other studies, which also point out that missing sleep takes points off children's IQ.

Stay awake only if you want the common cold. One hundred and 53 healthy men and women (aged 21–55) volunteered to report their sleep duration and sleep efficiency (percentage of time in bed actually asleep) for the previous night, and whether they felt rested to a team of American researchers led by Sheldon Cohen. After calculating average scores for sleep variables for 14 consecutive days, the participants were administered nasal drops containing a rhinovirus and monitored for development of a clinical cold. The results: lack of sleep makes you more susceptible to the common cold

Take a nap

After lunch, forget caffeine, take your shoes off, sit down comfortably and learn something from Charlie Brown's lovable dog Snoopy: take a short nap (in one of Charles Schultz's *Peanuts* cartoon strips Snoopy is contemplating: 'Learn from yesterday ... Live for today ... Look for tomorrow ... Rest this afternoon). Plenty of studies have established that naps of 10, 20 and 30 minutes all boost brain power (5 minutes is too short for a worthwhile nap). 'When all else fails, take a nap,' so goes an old proverb. Shut your eyes and relax. A 30-minute nap gives all the benefits of REM sleep.

Simply think you have slept well

Sure, you know about placebo effect. A patient's expectations and beliefs can greatly change the course of an illness, and medical researchers now see placebos as a key to understanding how the brain promotes faster healing. Placebo pills fool the mind into thinking that the problem is being taken care of. Similarly, 'placebo sleep' is like being told you're getting enough sleep. Researchers from Colorado College in the US devised an elaborate ruse in which 164 participating students were told that a new technique—which didn't actually exist— could measure their sleep quality from the night before. After hooking them up to fake machine, some were told their REM sleep from the night before had been above average, a sign that they were mentally alert. Others were told their REM sleep had been below average. Students in both groups got a five-minute lesson on sleep quality and its importance on mental abilities. The students who thought they got a night's sleep performed significantly better on real tests that assessed their ability to listen and process information. That's the power of positive sleeping.

Forget about 'alpha consciousness'

Alpha waves, which have a frequency between 8 to 13 hertz, are associated with relaxed mind or daydreaming; they are easily produced in the brain when quietly sitting in a relaxed position with eyes closed. New Age hucksters encourage people to undergo brain-wave biofeedback using commercially available devices to increase their production of alpha waves. It's true that people tend to display heightened proportion of alpha and theta waves while meditating or relaxing deeply. But there is absolutely no evidence that by boosting your alpha waves you can achieve deeper sense of consciousness and relaxation. It's cheaper and easier to take a nap, instead.

Some scientific estimates show that people are getting between one and two hours of less sleep a night than 60 years ago, which is leading to serious health problems. You can make your life better if you follow *The Time Traveler's Wife* (Audrey Niffenegger): 'Sleep is my lover now, my forgetting, my opiate, my oblivion.' Make it yours too.

Chapter 28

The power of expectation

learn our expectations influence our daily lives/*unlearn* to dismiss the placebo effect, which works on our expectations

If you expect nothing from anybody, you're never disappointed.

Sylvia Plath, *The Bell Jar*

It's an experiment you can try at home with your friends. Ask them to blind-test wines and then ask them to guess the price of the bottle from, say, from $10 to $100. You would be surprised to find out that your friends generally rate those wines higher that they thought were expensive. In realty, all wines are the cheapest ones, say around $10.

Baba Shiv, a Stanford University neuroeconomist who is an expert in the field of 'decision neuroscience', and colleagues conducted this experiment in a scientific way, rather neuroscientific way. They gave participants a taste of five different red wines (with prices marked as $5, $10, $35, $45 and $90) and then asked them to rate them. Unknown to participants the $5 and $45 wine and the $10 and $90 wine were identical. Participants, as expected, rated the expensive wines as more likable.

During the tasting sessions, the participants were hooked up to an fMRI machine. Their brain scans revealed that when tasting the 'expensive wines' the brain activity increased in areas of the brain known to be correlated with pleasantness of tastes, smells and even music. Other areas of brain associated with primary taste showed no change in activity. In a follow-up session after eight weeks, when wines were presented without prices, the participants reported no difference between the wines. The price was raising expectations and in a subtle way making a big difference.

In another experiment Shiv gave an energy drink thought to increase mental acuity to two groups of participants: one group paid the full price for the drink, the other was given a discount. After participants drank up, they were asked to solve a series of word puzzles. The result? Participants who bought the drink at a discount solved on average 30 per cent fewer puzzles than those who had paid the full price for the same drink. The results were within the same range experiment after experiment. 'It turns out you end up becoming dumber if you buy the product at a discount,' says Shiv.

It's our conscious belief that is equating price with quality. This belief works even though we know on some level that it's not always true. Our expectations mediate this placebo effect—the effect when a suggestion or belief that something is helpful actually makes it to become helpful. In medicine, the placebo (from Latin *placere*, 'to please') is typically a pill that looks and tastes like the drug but doesn't contain any drug.

Scientists sometimes use dummy pills to test the effect of a new drug. They randomly divide the patients in three groups: first group is given the drug being tested; the second gets no treatment; and the third the placebo. The no-treatment and placebo groups are known as control groups, the former shows how many patients would likely to get better by themselves and the latter shows the effect of belief in the drug (even if it's a sugar pill but patients do not know that). The results are always surprising: the placebo group may have zero to 100 per cent recovery rate, and the results are not clearly related to individual variables such as gender, age and culture (surprisingly, some trials have shown variation in results from country to country). The placebo effect depends on so many variables to have a simple relationship with anyone of them.

Trials by neuroscientist Predrag Petrovic of the Karolinska Institutet in Sweden have shown that placebo can also bring emotional relief. He first showed volunteers unpleasant pictures that included mutilated bodies. He then gave them an anti-anxiety drug to reduce their unpleasant perceptions of pictures. As our expectation plays a big role in the effectiveness of placebos, this step was to induce this expectation. The next day he gave them saline solution as a placebo but told that they were receiving the same drug. When they looked at the pictures again their unpleasant feeling was reduced by 29 per cent. Their brain scans revealed that placebo had reduced activity in the brain's emotional centre. Also importantly, the placebo increased brain activity in the brain areas that showed increased activity when placebo had been used to relieve pain. Those volunteers who expected the largest effect showed the largest changes.

Placebos don't work if you are hostile towards them. Researchers at the University of Michigan administered standard personality test to 50 healthy volunteers. The volunteers then received painful injections, followed by a sham painkiller—a placebo—to ease the resulting pain. The researchers found that the placebos worked better for volunteers with agreeable personality. 'Personality traits like straightforwardness and altruism are part of an overall capacity to be open to new experiences and integrate that information in a positive fashion,' says Kar Zubieta, the lead researcher. 'That probably drives the placebo effect.'

Placebos work even when people know that they are being given placebos. In a study, 80 patients with irritable bowel syndrome were instructed to take two sugar pills daily. After three weeks, 60 per cent of them reported improvement in the condition. If you have got a headache reading so much about the pleasing power of placebo, why bother taking a pain reliever? Suck a sugar pill, be nice to everyone around you and *believe* in the pain-relieving power of sugar pills. Would it work? Let me know if it does.

'The placebo changes what we expect,' Petrovic says. 'When we expect that something unpleasant should become less pleasant, it really does.' The placebo effect also works the other way around: when we think something pleasant would become unpleasant, it does, too. Many studies have shown that both positive and negative expectations determine how well a drug will work. The effects of negative thinking are more pronounced in patients with chronic medical conditions because they are more likely to have experienced years of frustration with ineffective medications. Not to worry, research tells us that 80 per cent of the population holds optimistic view of the future.

Tali Sharot, a UK neurology researcher and author of *The Optimism Bias*, has identified brain regions that fuel the brain's predilection for the positive. These areas are in the prefrontal cortex, where conscious decision-making takes place. Sharot warns that being overly optimistic has consequence, too, preventing us taking some precautions to avoid harm or misfortune.

Our expectations can also produce bad effects. Described as placebo's evil twin, the nocebo effect occurs when people claim to feel worse after taking placebo pills. Their headaches, fatigue, insomnia, stomach aches, nausea, dizziness and other symptoms are not in the mind; they are physical effects and can have a long-lasting impact on health. Studies show that a patient who expects to suffer painful symptoms is more likely to. When people think they are sick, they get sick. Researchers noticed the first-large scale nocebo (from Latin *nocere*, 'to harm') effect in the late 1990s when they came across an unusual finding: women, with similar risk factors, were four times more likely to die if they believed they were prone to heart disease. The higher risk of death had no underlying medical cause such as age, blood pressure, cholesterol or weight. It seems that patients can create their own nocebo effect unwittingly. Women report nocebo responses to therapy more than men do but, like the placebo effect, the nocebo effect is not clearly related to individual variables such as gender, age and culture.

The last word on placebos goes to English theoretical psychologist Nicholas Humphrey. He writes in *New Scientist* that societies are brimming with fake messages that have a powerful placebo effect, too: 'As human culture evolved, our ancestors were fortunate to discover that certain kinds of fake signals—pure make-believe—can do the job as well as the real things. Sometimes they can even do better.' His list of fake signals include, national anthems, rain dances, pyramids, gladiatorial shows, royal families ... and especially religious myths and rituals.

'Humans need... *fantasies* to make life bearable.' (Terry Pratchett, *Hogfather*). Make-believe signals packed as pink pills also make life bearable.

Chapter 29

Scarcity absorbs the mind

learn having too little time or money leads us to making bad choices/*unlearn* dieting, which is another kind of scarcity, has only physiological effects, not psychological

> It is deliberate policy to keep even the favoured groups somewhere near the brink of hardship, because a general state of scarcity increases the importance of small privileges and thus magnifies the distinctions between one group and another.
>
> George Orwell, *1984*

In *1984*'s dystopian society of the future Big Brother used scarcity to control the minds of people. In our real society, numerous studies prove that scarcity places demands on our mental capacity to make good decisions.

In one of the first such studies in the 1970s, psychologist Stephen West found that students at Florida State University, like most college students, rated their cafeteria food unsatisfactory. But their opinion changed dramatically just nine days later when they learned that because of a fire cafeteria meals would not be available for several weeks. The dull cafeteria food become more delicious the moment students realised it was less available. A perceived scarcity had affected the opinion of students.

Four decades later, economist Sendhil Mullainathan of Harvard University and psychologist Eldar Shafir of Princeton University conducted a series of laboratory experiments to discover the consequences of having too little. The participants were given resources that they could use to rewards while playing games. Participants who were randomly assigned to be 'poor' were given fewer resources, while the participants assigned to be 'rich' were given greater resources. When they were given the ability to borrow at exorbitant interest rates 'poor' participants borrowed heavily focusing on the need at hand, rather than the long term. This was not true for the 'rich' participants. It shows that scarcity promotes choices that people with less scarcity would know to avoid.

In another experiment the researchers asked shoppers at a US shopping mall to imagine their car required a repair costing $300. The shoppers, whose annual incomes ranged from $20,000 to $70,000, were free to pay for repair now, take out a loan to cover the cost of repairs or simply ignore it. They shoppers also completed a series of cognitive tasks that measured abilities such as logical thinking and problem solving. Performance on these tasks was the same regardless of income level. But when the researchers raised the cost of repair to $3000, the cognitive abilities of shoppers at the lower income levels declined by 40 per cent.

Both experiments show that scarcity impairs people's ability to think clearly. Scarcity, a broader term than poverty, is an involuntary preoccupation with an unmet need, such as a shortage of money or time. It can capture our attention and impede our ability to focus on other things. Why? When we worry, even unconsciously, about not having enough money, the worrying end up consuming some of our cognitive resources. Chronic scarcity can cause chronic stress that can keep people at a low socioeconomic status by impairing their cognitive abilities. It even makes the body pump out more stress hormones, which ravage the immune system. In the brain, prefrontal cortex, the decision-making region, and amygdala, the emotional centre, are quite sensitive to stress. Brain-imaging studies show that people who have lived under poverty and long-term stress caused by it show less ability to regulate amygdala. Poor childhood has also been associated with a lower working memory capacity in young adulthood. More stress led to even lower capacity.

In their book, *Scarcity: Why Having Too Little Means So Much*, Mullainathan and Shafir ask readers to imagine how they might pack a suitably large suitcase for a trip. Now, they say, imagine instead packing a small suitcase for the same trip. Packing the small suitcase forces trade-offs. Both the large and small suitcases require a choice of what to pack and what to leave out. Yet psychologically only the small suitcase really feels like a problem. Understanding the differences in how we pack is crucial for understanding, the authors say, how scarcity creates more scarcity. Scarcity controls the mind: when we lack money, time, food or friends, it consumes us, making us 'dumber' as we are unable to think of anything else. Good long-term decisions require cognitive resources. Scarcity leaves far less of these resources at our disposal.

Mullainathan also applies his studies of scarcity to dieting. 'Diets don't just reduce weight, they can reduce mental capacity,' he says. 'In other words, dieting can make you dumber.' Logical and spatial reasoning, self-control, problem solving, and absorption and retention of new information, together these tasks measure 'bandwidth' which in this context means mental capacity.

In a study by the University of Toronto psychologist Janet Polivy young women were told that they weighed five pounds more, or less, than their real weight. The dieters who heard they were heavier reported lower self-esteem and more negative moods than the non-dieters who received the same information. These depressed dieters ate significantly more in a staged 'taste test'. Dieting is a personality variable, not just a behaviour restricted to eating, says Polivy. Mullainathan agrees when he says: 'Diets do not just strain bandwidth because they leave us hungry. They have psychological, not just physiological, effects.'

We tax our bandwidths when we feel we have too little time or too little money. Bandwidth scarcity has far-reaching consequences, says Mullainathan. 'We all use bandwidths to make decisions at work, to resist the urge to yell at our children when they annoy us, or even to focus on conversation during dinner or in a meeting.' Scarcity robs us of good decision-making abilities.

University of Westminster social psychologist Viren Swami has even found an unlikely outcome of scarcity: who you fancy can change depending on how hungry you are. He says that scarcity of food affects preference for body weight: 30 hungry male participants preferred figures with a higher body weight and rated as more attractive heavier figures than 31 satiated male participants. Moral: Never go out hungry on your first date.

How do we avoid effects of scarcity on our minds? Here's some tips for managing scarcity: (1) focus your mind on what is good about life; (2) avoid comparing yourself with others; (3) stop obsessing—take a walk, call a friend or read a book to break the negative emotional cycle; (3) to manage finances take measures such as not taking your credit card to the shopping mall and signing up for automatic bill payments; (5) don't be greedy—when resources are scarce people become competitive; (6) do not miss quality time with family and friends as loneliness is a social form of scarcity; (7) be health conscious while shopping rather than at every meal time to free up your cognitive bandwidth; and (8) regular exercise and sleep are more important to maximise mental bandwidth rather than hours worked.

There are times when scarcity can be good. When you have less money or time, it may result in a 'focus dividend' in the form of increased activity. Scarcity can also result in 'tunnelling': you focus on one thing so much that you neglect something else. 'Focus is positive: scarcity focuses on what seems, at that moment, to matter most,' Mullainathan and Shafir write in their book. 'Tunnelling is not: scarcity leads us to tunnel and neglect other, possibly more important, things.'

Until our world is 'scarcity-proof' we all need both logic and kindness:

> 'But still,' Ayumi said, 'it seems to me that this world has a serious shortage of both logic and kindness.'
> 'You may be right,' Aomame said. 'But it's too late to trade it in for another one.'
>
> Haruki Murakami, *1Q84*

Chapter 30

Let your body do the talking

learn how your body language can influence your personality and how you feel/*unlearn* to be smart about interpreting body language of others as its rules are 'arbitrary' and meaning can be lost in translation

> The trouble with England, he thinks, is that it's so poor in gesture. We shall have to develop a hand signal for 'Back off, our prince is fucking this man's daughter.' He is surprised that the Italians have not done it. Though perhaps they have, and he just never caught on.
>
> – 'He' is Thomas Cromwell, the novel's protagonist
> Hilary Mantel, *Wolf Hall*

Even if you speak only English, you're bilingual. You speak English and body language. Your body language is, in many ways, much richer and older than your words.

One-finger salute—the middle finger extended with the other finger held beneath the thumb—is one of the most ancient insult-and-inflame gestures known to us. 'The middle finger is the penis and the curled fingers on either side are the testicles,' says Desmond Morris, an English zoologist and author of *The Naked Ape*. 'By doing it, you are offering someone a phallic gesture. It is saying, 'this is a phallus' that you're offering to people, which is a very primeval display.' Even civilised ancient Greeks didn't hesitate to give their fellow citizens the finger. A character in *The Clouds*, a comedy written in 423 BC by the comic poet Aristophanes, gestures with his middle finger ('What is it then, other than this finger here?').

Popular culture is full of insights into body language. Do we really betray ourselves through body language? Scratching nose, looking up and to the right and averting gaze are perceived as signals that the person is lying, but there is no evidence to support this myth. The meanings we assign to postures such as swaggering walk (confidence), arms crossed (defensive) and palms up when talking (trustworthy) are not necessarily true, either. Says Nicholas Epley, an American behavioural scientist and author of *Mindwise: How We Understand What Others Think, Believe, Feel, and Want*: 'There is an illusion of insight that comes from looking at a person's body. Body language speaks to us, but only in whispers.'

The whispers are not true when people claim that a person looking up to their right suggests a lie whereas looking up to their left indicate truth telling. After observing the eye movements of volunteers telling lies in laboratory experiments, a team of UK psychologists led by Richard Wiseman has found no evidence for this claim. Other studies show that body whispers are true when we associate fidgeting with embarrassment, hands on hips and wide stance with power and confidence, and raised arms and chin up with triumph and pride.

A comprehensive analysis of the accuracy of deception detection made by more than 24,000 people who had participated in 206 studies exposes that, on average, people were only correct about determining when people were lying 54 per cent of the time. 'They would have been right 50 per cent of the time just by guessing,' comments American psychologist Bella DePaulo who co-wrote the review article with Charlie Bond. Their analysis, however, showed that liars, on average, appear more nervous than truth-tellers. 'But here's the catch: not all people appear more nervous when they are lying than when they are telling the truth,' she says. Her verdict on the accuracy of reading body language is terse: 'The sign are not there, even for the people genuinely interested in seeing them.' Epley's own research has also come to a similar conclusion: 'Reading people's expressions can give you a little information, but you get so much more by talking to them. The mind comes through the mouth.'

Even if the 'mind comes through the mouth', facial expressions are crucial for social communication and our brains are attuned to respond quickly to them. It takes a person only about 100 milliseconds to determine whether they like or don't like another person based on their face. If you take even longer to look at the face of a Japanese person, you may not find out how the person is feeling. You are advised to pay attention to the person's tone, not the face as Japanese are taught to mask overt display of emotions. 'I think Japanese people tend to hide their negative emotions by smiling, but it's more difficult to hide negative emotions in the voice,' explains Akihiro Tanaka of Waseda Institute of Advanced Study in Japan. Understanding emotional vocal cues are as important as reading someone's body language.

You may or may not be an expert in reading body language, but your own posture has a big impact on your own body and mind. Holding an open and expansive pose for a few minutes can actually change your hormones and behaviour just as if you had real power. 'It can increase the dominance hormone testosterone and decrease the stress hormone cortisol,' says psychologist Dana Carney of the University of California who has led a study in which participants were guided for two minutes into either high-power poses (standing tall with legs apart and hands on hips, for example) or low-power-postures (slumping or leaning back with arms or ankles crossed). Afterwards, the participants played a gambling game where the odds of winning were 50:50. Some 86 per cent of high-power posers risked losing compared with 60 per cent of the low-power posers. The willingness of high-power posers to gamble was also linked to increase in testosterone and decrease in cortisol, while the low-power posers showed a decrease in testosterone and increase in cortisol.

'Postures aren't just an expression of how we feel,' Carney says. 'They can also inform the brain by changing our physiology.' Before a stressful event such as an interview, she recommends the ways you stand and sit to improve your chances of success: (1) Open your torso by putting your hands on your hips, and take a wide stance to expand your lower body; you can do this in the elevator on the way to an important meeting. (2) While waiting, sit with one hand around another chair on the outside of the armrest to expand your chest; do not cross your legs too tightly.

We make various gestures when explaining something, but we also move hands when simply talking. Even congenitally blind people move hands when talking suggesting that hand gestures are not for others but for ourselves. Susan Goldin-Meadow, a psychology professor at the University of Chicago and author of *Hearing Gesture: How Our Hands Help Us Think*, who has conducted many studies of the importance of hand gestures to learning, thinks gestures reveal subconscious. 'We change our minds by moving our hands,' she says. To her, body movements are also part of how we learn—gestures are involved not only in processing old ideas, but creating new ones.

In one of her studies she found that children who were asked to gesture while learning algebra remembered three times more than their classmates who did not gesture. At times gestures can also mislead children. She describes a classroom situation when a teacher's gestures inadvertently led a student to an incorrect strategy for solving an algebra problem. 'While verbally describing the correct strategy, the teacher pointed out all four numbers in the equation, a gesture the child read as an instruction to add up all the numbers in the problem,' she says. 'The child then put that sum in the blank.' Hand movements can mislead as well as inform. Yet, she firmly believes, in general, how teachers and children move their hands in the classroom 'can smooth the path toward knowledge'.

Should you follow the old advice 'stay still while speaking'? Oh, no, don't torture your body—and mind. Anyway, the words will still emerge from your body.

> The words emerge from her body without her realizing it, as if she were being visited by the memory of a language long forsaken.
>
> Marguerite Duras, *Summer Rain*

Chapter 31

Laughter is sunshine

learn how your sense of humour can improve your health and creativity/*unlearn* funniest people are always the healthiest

'Laughter is poison to fear.'

– Catelyn to Robb
George R.R. Martin, *A Game of Thrones*

Laughter is something humans have been enjoying for millions of years, regardless of race, culture and language. It's something you have been enjoying since you were about 14 to 16 weeks old. It's also the simplest and the fastest way to relax. Given the choice between laughter and a pill to relax, you're better off with a bit of ha-ha-ha or ho-ho-ho. Your attempt at ha-ho-ha-ho would not amuse the world's foremost expert on laughter, Robert Provine, an American neurobiologist. 'Try to stimulate ha-ho-ha-ho laugh—it should feel quite unnatural,' he says. 'When there are variations in the notes, they must often involve the first or last note in a sequence.' Go for a bit of cha-ha-ha or ha-ha-ho and he would sure join you. For these are the possible variants of laughter.

No one likes to use old-fashioned clichés but 'laughter is the best medicine' is worth mentioning here. A slew of research shows that laughter is indeed therapeutic, mentally and physically. Laughter gives our brains a rush of endorphins, the hormones that are responsible for our general sense of wellbeing. Just after five minutes of laughter you'll start feeling really great.

To find out what goes inside our brains when we laugh a team of scientists led by Dirk Wildgruber, professor of neuropsychiatry at Eberhard Karls University of Tübingenin in Germany, scanned the brain activity of 18 young men and women as they listened to recorded laughter. The laughter was recorded in three situations: feeling joy, taunting someone and being tickled. The volunteers were asked to identify the type of laughter. In most cases they correctly identified the two types of social laughter but they were slightly less accurate at correctly labelling ticklish laugher. (Interestingly, most people, especially adults, say they hate to be tickled but still they laugh when tickled; and you do not laugh when you tickle yourself.)

After analysing 1200 brain images the German team found that there was a stronger connection between the brain areas involved in processing sound and vision, as well as those that allow us to 'mentalise' what other might be thinking and feeling. The team also found that when listening to taunting laughter, there was a stronger connection between sound-processing and 'mentalising' areas of the brain. This suggests the brain infers the social consequences of taunting laugher from acoustic signals. During joyous laughter the visual area was more active.

Other brain-imaging studies show that laughter is truly infectious. TV producers of comedy shows have long known that laugh tracks do increase audience laughter. But in the absence of a joke, the laughter track itself cannot evoke laughter. The most dramatic episode of the infectious power of laugher happened in 1962 in Tanganyika (now Tanzania). Three schoolgirls in a group of 12- to 18-year-olds couldn't stop laughing. Soon their uncontrollable guffaw spread to the whole school. It then propagated from school to school to village to village. Laughing attacks lasted from a few minutes up to a few hours. The epidemic was so severe that the authorities closed down schools until the epidemic died down after six months. About a thousand people had been struck down by the 'laughing disease' as the local doctors described it. They couldn't find any explanation for it.

Fast forward to a team of cognitive neuroscientists at the University College London who know why laughter is contagious. They played a series of sounds to volunteers while measuring their brain's response using an fMRI scanner. Some of the sounds were positive, such as laughter or triumph, while others were unpleasant, such as screaming or retching. All of the sounds triggered a response in the volunteer's brain in a region that prepares the muscles in the face to respond accordingly, though the response was greater for positive sounds, suggesting that these were more contagious than negative sounds. 'This response in the brain, automatically priming us to smile or laugh, provides a way of mirroring the behaviour of others, something which helps us interact socially,' says Sophie Scott, the lead researcher. It seems that the old saying 'laugh and the whole world laughs with you' is true after all. Sharing a laugh brings people closer together.

In a study by James Rotton of Florida International University people who watched funny movies after surgery requested 25 per cent less pain medication. Another study showed that watching an episode of TV show *Friends* reduced anxiety three times as effectively as just sitting and resting. When you're laughing, as mentioned above, the brain releases endorphins, which raise our ability to ignore pain. This pain-relieving chemical is also created in response to exercise, massage, excitement, spicy food, love and sexual orgasm, among other things. Laughter excites abdominal muscles, which, in turn, trigger the release of endorphins. As endorphins are also released during exercise, laughter is as effective as exercise.

So, go ahead, don't text but truly LOL. If you would like to twitter, here's a titbit: when seeking a partner, men prefer women who laugh at their jokes, whereas women prefer men who can make them laugh. More titbits: (a) women laugh more than men; (b) contrary to popular belief, men are not funnier than women; and (c) deaf signers also interject their signing with laugher—laughter involves higher order cognitive processes rather than lower level motor processes that produce voice.

Waiting for that eureka moment of sudden inspiration? Aha! Take a break to watch a funny show or movie. Laughter would boost your mood and lead to that long-sought insight. Brain-imaging studies show that people in a positive mood have more activity in a region of the brain called anterior cingulate cortex (ACC). When we have a sudden insight ACC also lights up. Therefore, happy mood helps prepare the brain to find novel solutions to problems. If you watch an anxiety-producing show or movie, activity in ACC decreases which, in turn, diminishes creativity.

Sadly, this happy story of laughter has a not-so-funny ending. Research shows that people with great sense of humour tend to engage in less healthy lifestyles. A long-term study of Finnish police officers found that those who were seen as funniest smoked more, weighed more and were at greater risk of cardiovascular disease than their peers. Do funny people, having a generally less serious perspective, take health risks more lightly and engage in more risky behaviours than less humorous people?

After analysing data from 404 men and 119 women comedians from UK, US and Australia who had completed a specially devised questionnaire, Oxford University psychologists found that those working in comedy may be more disposed to 'high levels of psychotic personality traits'. The comedians scored particularly high on personality traits such as being unsociable and depressive as well as more extrovert manic-like traits.

Funny people might not always be having much fun, but a review of 'humour and ageing' studies by anthropologist Gil Greengross of the University of New Mexico has some good news: older people enjoy humour more than younger people, even if they have increasing difficulty in understanding jokes. However, the actual amount of laughter declines with age.

'Laughter chases winter from the human face' (Victor Hugo, *Les Misérables*). Every season has a reason for laughing out loudly. Ha-ha-ha.

Chapter 32

Step into your shoes before you step into someone else's shoes

learn the importance of empathy and self-compassion/*unlearn* empathy is unchangeable and has no dark side

> Not even one's own pain weighs so heavy as the pain one feels with someone, for someone, a pain intensified by the imagination and prolonged by a hundred echoes. He kept warning himself not to give in to compassion, and compassion listened with bowed head and a seemingly guilty conscience.
>
> – 'He' is the novel's protagonist Tomas, a surgeon
> Milan Kundera, *The Unbearable Lightness of Being*

When you say to someone 'I'm sorry about you being in pain', you are showing sympathy to that person, but when you say 'I feel your pain', you are showing empathy. Empathy makes you sensitive to the feelings and thoughts of others. It's like stepping into someone's shoes and then stepping out of it.

Compassion is an emotional response when you perceive that someone is suffering and there is a genuine desire to help. While we are going through a dictionary, let's look up another word: altruism, an action that benefits someone else; it may or may not involve empathy or compassion. Compassion, however, is accompanied by empathy and altruism.

Feeling what another is feeling

In a way, empathy encourages us to enter imaginatively into the lives of others. Our brains do that imaginative work through mirror neurons: a type of neuron that respond equally when we do something and when we see someone else doing the same thing. In the 1990s neuroscientists made a surprising discovery when they recorded brain activity of monkeys performing certain tasks. They found that individual neurons would only respond to very specific actions; for example, when pushing a button or pulling a lever. They noticed something more surprising: when the monkey reached for a peanut a certain neuron would fire. Amazingly, when the monkey saw another monkey reach for the peanut, the same neuron fired again. The neuron seemed to be 'mind-reading' intentions of another's actions; it was doing a kind of internal virtual reality simulation of the other monkey's action to figure out what he was 'up to'. The actions of mirror neurons are involuntary and automatic. We don't have to think about what other people are doing or feeling, we just know it.

The question that mirror neurons give us the ability to read the intentions of other is debatable. Empathy may or may not be hardwired, but we do have some innate capacity for empathy. When you see a child jamming their finger in the door you immediately tense up and your 'ouch' appears almost simultaneously as the child's. This capacity to be concerned about human beings others than ourselves extends even to people living on the other side of the world.

When you read literature on empathy the name of Kathy Ficus pops up regularly. No, she wasn't a great thinker on the subject but a three-year old girl who fell down a narrow shaft of an abandoned well in San Marino, California in 1949. She was trapped 90 feet (27 metres) underground, in a pipe just 14 inches (36 centimetres) wide. The rescue effort transfixed the whole nation for more than 27 hours. 'A hard-boiled city poured out its tears and silent prayers as frantic men worked in a dark tunnel to rescue the little girl,' *Los Angeles Times* noted. When the news of her death came, people glued to their radios or black-and-white TVs felt the pang of sorrow. Cynics dismiss this phenomenon of mass empathy simply as emotional contagion: millions of people concerned about one child in trouble, while ignoring millions of others who are starving. Yet, the immense power of empathy has been demonstrated again and again in other similar incidents.

Empathy directs us towards moral action, but it mostly fails when that action demands some personal sacrifices. Jesse

Prinz, a philosopher at City University of New York agrees that empathy is a good thing from a moral point of view but challenges the notion that somehow empathy is necessary for moral conduct. He writes in a paper 'Is empathy necessary for morality?': 'One might hold the view that empathy is necessary for making moral judgment. One might think empathy is necessary for moral development. And one might think empathy is necessary for motivating moral conduct.' He rejects each of these conjectures and concludes, 'Its contribution is negligible in children, modest in adults, and non-existent when costs are significant.' Paul Bloom presents a succinct counterargument in *The New Yorker* magazine: 'Empathy is what makes us human; it's what makes us both subjects and objects of moral concern. Empathy betrays us only when we take it as a moral guide.'

The darker side of empathy shadows us when we show the most willingness to sacrifice ourselves for those we care about most: our family, relatives and friends, and social and religious groups we belong to. American psychologists Daryl Cameron and Keith Payne have studied a phenomenon that call 'the collapse of compassion': When they measured people's emotional experiences in real time—rather than their predictions—they found that rather than feeling more compassion when more people are suffering, people ironically feel less. In other words, they purposefully restricted their empathy when empathising could prove costly.

Several other studies show that wealth reduces empathy and compassion. In an experiment, participants watched a video of a child suffering from cancer; the participants' social class was measured by the level of their income and education. The results showed that the participants at the lower end reported feeling more compassion while watching the video. Their heart rate also slowed down while watching the video—a response associated with paying attention to the feelings of others.

A study of American middle and high school students across a wide spectrum of races, cultures and classes reveals that students rank achievement or happiness over caring for others. The 2014 study 'The Children We Mean to Raise: The Real Message Adults are Sending About Values' by Harvard University's Graduate School of Education points out that about 8000 of the 10,000 students who participated in the study reported that their parents and teachers 'are more concerned about achievement or happiness than caring for others'. It's ironic that focus on achievement and happiness in affluent communities, the study says, doesn't appear to increase either children's achievement or happiness.

If you think, for whatever reason, your empathy needs upgrading, there is help at hand. All you need to do is read literary fiction. New York social psychologists David Kidd and Emanuele Castano conducted five studies in which they asked participants to read 10 to 15 pages of either literary fiction, popular fiction or non-fiction. The participants who read literary fiction performed better on subsequent tests measuring empathy, social perception and emotional intelligence than those who read popular fiction and non-fiction. Literary fiction focuses on psychology of characters and their relationships. Navigating these ambiguous fictional worlds filled with emotional nuance and complexity serves us well in life by enhance our ability to understand someone else's mental state.

For those who are interested, Kidd and Castano's reading list included: literary fiction (*Chameleon* by Anton Chekhov, *The Runner* by Don DeLillo, *Round House* by Louise Erdrich, and *Blind Date* by Lydia Davis); popular fiction (*The Sins of the Mother* by Danielle Steel, and *Gone Girl* by Gillian Flynn); and non-fiction (articles such as 'How the Potato Changed the World' from *Smithsonian* magazine. Not everyone is impressed with their choice of reading list and methodology. You may continue reading *Fifty Shades of Grey* trilogy without any fear of loss of your empathy for Anastasia Steele or Christian Grey.

You can still increase your empathy by practising 'compassion meditation' which focuses on a specific person while repeating a phrase like 'May you be free from suffering'. In a study conducted by a team of eight researchers led by cognitive neuroscientist Helen Weng of the University of Wisconsin–Madison, the participants concentrated on a loved one, a friend, themselves, a stranger and then someone they were in conflict with. Another group of participants performed general positive thinking. The participants who did compassion meditation were more likely to spend their money to help than those who did positive thinking. Their brain scans also showed increased activity in areas involved in empathy. Like muscles, you can build up empathy by exercise.

Feeling for yourself

The concept of compassion has an important place in Buddhism. 'Compassion,' says the Dalai Lama, 'is an aspiration, a state of mind, wanting others to be free from suffering.' In Buddhism, self-compassion is the wish to free ourselves from suffering. When psychologists talk about self-compassion, they stress kindness towards our failings.

Self-compassion is having compassion for yourself: caring for yourself, treating yourself kindly, enabling yourself to let painful emotions pass.

It has three components: (1) Showing kindness towards yourself when facing pain or failure—treat yourself as you would treat your friends and family. (2) Seeing your experiences as part of a larger human experience—when you criticise or judge yourself, think 'we are all imperfect'. (3) Holding painful thoughts and feeling in balanced awareness—let your painful emotions pass, compassionately and with kindness.

Kristin Neff, a pioneer in the field of self-compassion research and author of *Self-Compassion: The Proven Power of Being Kind to Yourself*, has drawn on the writings of Buddhist scholars. 'The biggest reason more people aren't more self-compassionate is that they are afraid they'll become self-indulgent,' she says. 'They believe self-criticism is what keeps them in line. Most people have gotten it wrong because our culture says being hard on yourself is the way to be.'

Research shows that self-compassion, not self-criticism, helps us in many ways. Neff and her colleagues have found that even a little self-compassion may influence eating habits. At the beginning of their study participants were asked to eat doughnuts. One group, however, was given a message on self-compassion with the food, such as 'Don't be hard on yourself, everyone in the study eats them all the time'. Later that day, participants were given the chance to eat sweets. Those who were not given the message on self-compassion ate more. The results show that participants who felt bad about eating doughnuts engaged in 'emotional' eating, while the participants who gave themselves permission to enjoy the sweets didn't overeat.

Self-compassion helps people to survive break-up without bitterness and get back them on their feet. This is the finding of a study led by psychologist David Sbarra of the University of Arizona. The study involved 105 men and women with an average age of 40 who had been married for more than 13 years and divorced an average of three to four months. On the first visit, the participants were asked to think of their former partner for 30 seconds, and then talk for four minutes about their feelings and thoughts related to the separation. The researchers listened to the audio files and rated participants' level of self-compassion using a standard measure. The participants, who were also assessed on other psychological traits, were asked to report three times over nine months on their adjustments, including how often they had negative thoughts and emotions about their former partners. As expected, the participants who had high level of self-compassion at the start were good at coping with their broken heart and recovered faster. 'You can't change your personality so easily,' says Sbarra, 'What you can change is your stance with respect to your experience. Understanding your loss as part of bigger human experience helps assuage feelings of isolation.'

Here're three simple ways to boost your self-compassion: (a) be your own best friend; (b) you are not alone, you're part of a common humanity; and (c) take time out for quiet thinking, contemplation or meditation.

The diametrically opposite of self-compassion is self-loathing. In *Unworthy: How to Stop Hating Yourself*, Anneli Rufus, an American journalist, describes her own and others experiences of self-loathing. 'Some people think that they're really stupid; some people think they are weak; some people think they're ugly,' she says. 'It colours their daily life. You dress as fast as possible to hide your body without looking at yourself in the mirror because you can't stand it.'

How can self-loathing people help themselves? Her advice is to break those habits that keep us rooted in our self-loathing: 'Do you really feel that sorry about everything you do? Do you really feel like you need to beg everyone for permission? Probably not. Look at the things you do as if they were on a movie screen and take away the *I did because I'm an idiot*. The more you become aware of that thinking and those habits, the easier it becomes to shift them.'

'When we don't know who to hate, we hate ourselves.' (Chuck Palahniuk, *Invisible Monsters*). Empathy and self-compassion help us erase hatred from our minds for others and ourselves.

Chapter 33

Let your body do the thinking

learn how your senses and body movement influence your thoughts/*unlearn* your thinking is done by your brain alone

'I move, therefore I am.'

– the novel's heroine, a slender assassin named Aomame

Haruki Murakami, *1Q84*

You are holding a cup of coffee and someone approaches you to make a small donation to a worthy charity. Would it make a difference to the amount you are likely to donate—or not donate at all—if the cup contained hot or iced coffee? Before you answer, think of the words 'warm' and 'cold', which are used in everyday life to describe abstract concepts. A 'warm smile' is synonymous with kindness, and 'cold eyes' suggest a distant or selfish personality. We won't go as far as 'cold-blooded murder', but would let pass an 'icy stare' or a 'cold shoulder'.

A few years ago Lawrence William, a psychologist at the University of Colorado, asked undergraduate students to hold their cups of coffee, either hot or iced, and then fill out a personality-impression questionnaire. In the questionnaire the students were asked to rate 10 personality traits of a fictional person after reading a brief description of the person. The students weren't aware that holding the cup was part of the experiment, but the effect was quite meaningful and astonishing. The students who held hot cups judged the fictional person to be 'generous', 'caring' or 'sociable' than those who held cold cups.

In another experiment, William asked participants to hold hot or cold therapeutic pads for a moment and then judge the quality of a product. As a token of appreciation for their time, the participants could choose either a reward for themselves or a gift voucher for a friend. Fifty-four per cent of those who handled a hot pad, chose the voucher for a friend compared with only 25 per cent who held the cold pad.

The two experiments show that hot and cold sensations can influence our feelings, and it happens without our awareness of this influence. Other studies have shown: (1) Even holding a teddy bear makes you more ethical. (2) Wearing smart clothes makes you smarter: people who wore white lab coats made half as many mistakes on attention-related tasks as those wearing their regular clothes.

The idea that physical sensations evoke abstract ideas and shape basic perception is known as embodied cognition. For example, brain-imaging studies by Oshin Vartanian, a neuroscientist at the University of Toronto, to test people's reactions to pictures of household interiors, asking them to rate rooms as 'beautiful' or not 'beautiful' show that curved surfaces tend to make people feel relaxed and hopeful. Viewing curved spaces fires up neurons in anterior cingulate cortex, an area of the brain associated with aesthetic judgments. An earlier study by other researchers found sharp angles activate amygdala, the brain's fear centre. While you're waiting for your curved screen TV, learn a bit about ceilings before you start to refurnish your room: higher ceilings tend to encourage abstract thinking, while under lower ceilings people are more likely to think concretely.

Our brains do not make an artificial distinction between thinking and physical perceptions. For example, both the physical and psychological version of warmth information is processed in a brain region known as insular cortex, also called insula. This means the brain is part of a broader system that involves perception and action. The great 17th-century French mathematician and philosopher René Descartes was sceptical of almost everything, even his own existence. He lost this scepticism after reaching the conclusion, *'Cogito, ergo sum'* (*'I think, therefore I am'*). It's time to supplement philosophy's most famous statement with Aomame's *'I move, therefore I am'* or perhaps *'I move, therefore I think'*.

Art Glenberg, professor of psychology at Arizona State University and one of the founders of embodied cognition field, gives an example how the idea of embodied cognition can help children learn to read. When a child is reading a story about a farm and they read a sentence such as 'Then the farmer drives the tractor to the barn'. Reading is not simply being able to say the words. 'The idea of reading is being able to emulate or imagine the situation of the farmer getting into the tractor and driving the tractor to a particular location,' he says. 'The way we teach the children to do that is by having them read, but with toys available so that after they read a sentence like that they literally take the toy farmer and put the farmer into the toy tractor and move the toy tractor to the barn.' (The importance of hand gestures to learning explored in 'Let body do the talking' is also based on the idea of embodied cognition; and also the relationship between smiling and happiness in 'Turn up the wattage of your smile'.)

A quick look at some of other interesting findings in embodied cognition—the latest sexy topic, as some call it, in cognitive science.

Weight is indeed important. In your opinion, the idea that follows may not 'carry weight' (able to influence) or 'add weight' (place emphasis) to your thinking but we think it's 'worth its weight in gold' (useful, helpful or valuable). To test the hypothesis 'weight is embodiment of importance', psychologist Nils Jostmann of the University of Amsterdam and colleagues conducted four experiments. In one of their experiments the participants were randomly given light or heavy clipboards and asked to fill out a questionnaire to estimate the value of six foreign currencies. The participants who held heavy clipboards estimated the currencies to be more valuable than who held light clipboards. Their conclusion after three other similar tests: the abstract concept of importance is linked to our body's experience of weight.

Would reading a heavy tome make you smarter? 'Weigh' it carefully before you answer. It would certainly make you feel more important, if not smarter. The best way to become smarter is to lie down. Researchers at the Australian National University have found that people solved anagrams in about 10 per cent less time when lying down compared with standing. If you're a supine thinker, you know you are more relaxed on your back than your feet.

Physical touch influences our social judgments and decisions. Ponder as you hold this book: as you run your hand over it and if you find the texture rough, you may not like this book. If you find it soft and silky, you may like it. Joshua Ackerman, a psychologist at the Massachusetts Institute of Technology and colleagues conducted a series of experiments to probe the impact of texture and hardness on how we perceive the world. In one experiment, they asked participants either to arrange rough or smooth puzzles before hearing a story about a social interaction between two people that included an exchange of sharp words and some teasing. Those who worked with rough puzzles described the interaction in the story as more adversarial and harsh than those who worked with smooth puzzles. In a test of hardness, participants seated in hard or soft chairs engaged in mock haggling over the price of a new car. The people in hard wooden chairs were less willing to negotiate. We hope you are not sitting on a hard chair while reading this book.

The researchers say that touch appears to be the first sense we use to experience the world; for example, by equating the warm and gentle touch of our mother with comfort and safety. Eventually, the physical and the abstract become linked. Metaphors, such as a 'rough day' and 'hard done by', reflect the underlying links between our physical experience and our mental understanding.

Sprawling on a big chair can make you dishonest. Everyday our body postures are shaped by the furniture in our homes or workplaces and the seats in our cars. These physical environments directly lead to an increase in dishonesty, concludes a series of laboratory and field studies headed by American management expert Andy Yap. People standing in 'expansive' postures, either explicitly or inadvertently, were more likely to accept money they weren't owned. People working at a large desk space were more likely to cheat on a puzzle completion task than who had less work space. In a video-game driving simulation, people sitting in 'expansive' seats drove more recklessly and had more 'hit and runs' compared to people sitting in smaller seats. One field study revealed that drivers in more expansive driver's seat were more likely to be illegally parked on New York streets. 'Power causes you to focus on rewards and take risks to achieve those gains,' comments Yap.

Metaphors and creativity. By considering the problem 'on the one hand, then on the other' and then 'thinking outside the box' they easily 'put two and two together'. Nearly 400 college students put these metaphors to their metaphorical use: (a) to generate ideas while first holding out their right hand and then their left hand ('on the one hand, then on the other'); (b) to completing word tasks by either physically sitting inside or outside a box or engage in problem solving by walking in a rectangular path versus freely walking ('thinking outside the box'); and (c) to converging multiple ideas to find solutions while combining two objects ('putting two and two together').

The researchers Jeffrey Sanchez-Burks and Suntae Kim of the University of Michigan Ross School of Business report that physically and psychologically embodying creative metaphors promotes fluency, flexibility and originality in problem solving. 'The acts of alternately gesturing with each hand and of putting objects together may boost creative performance,' says Sanchez-Burks. 'Literally thinking outside or without physical constraints, such as walking outdoors or pacing around, may help eliminate unconscious mental barriers that restrict cognition.'

'You choose your behaviours based on their metaphorical resonances.' (John Green, *The Fault in Our Stars*). You're likely to choose the right behaviours if you let your mind rest inside the box and let your body do the thinking outside the box.

Chapter 34

Blowing hot and cold on self-esteem

learn how self-esteem affects your real and online life/*unlearn* high self-esteem is the fountain of all human goodness

> 'I have to admit it humbly, *mon cher compatriote*, I was always bursting with vanity. I, I, I is the refrain of my whole life, which could be heard in everything I said ... When I was concerned with others, I was so out of pure condescension, in utter freedom, and all the credit went to me: my self-esteem would go up a degree.'
>
> – the novel's protagonist Jean-Baptiste Clamence whose 'confession' isn't just his story—it's ours as well
>
> Albert Camus, *The Fall*

'Don't wrap your mind for ever round yourself,' says Aristotle in Aristophanes' comedy of ideas, *The Clouds*. Like Jean-Baptiste Clamence, Aristotle was not interested in pushing his self-esteem up by a degree or two. But if he was living today he would be trawling through Amazon's (at the time of the last count) 8,000 books on self-esteem, not to improve his self-esteem, but to find out why self-esteem is such a thriving business today. He would be fascinated by Western culture's preoccupation with personal happiness and self-esteem, in that order.

How high do you stand on the much prized self-esteem ladder? Take the test at the end of the story to find out your opinion of yourself. Or, log into Facebook to weigh your self-worth. Looking at your Facebook pages might even enhance your self-esteem. This is probably because Facebook allows you to choose what you want to reveal about yourself and filter anything that might reflect badly. Furthermore, feedback from friends posted on Facebook tends to be overwhelmingly positive. Some studies indicate that how we are able to present ourselves to others is important to self-esteem. By providing multiple opportunities for selective self-presentation—through photos, personal detail and witty comments—Facebook can influence impressions of the self.

But why so much fuss about impressions of the self? Does high self-esteem always translate into success? Or, does low self-esteem lead to various undesirable behaviours? Seems not.

Studies by well-known American psychologist Roy Baumeister show that people with high self-esteem think they make better impressions, have stronger friendships and have better romantic lives than other people, but evidence doesn't support their self-flattering views. 'If anything, people who love themselves too much sometimes annoy other people by their defensive or know-it-all attitudes.' he says. Self-esteem doesn't make adults perform better at their jobs either, he adds. His research also found that bullies, contrary to popular belief, do not suffer from low self-esteem. His advice to all those who are logged into Facebook to check flattering posts from their virtual friends to pep up their self-esteem: 'Forget about self-esteem and concentrate more on self-control and self-discipline.'

Similar messages come from other studies of self-esteem: high self-esteem has a few modest benefits, but it can produce problems and is most irrelevant to success. The way to go is to think less about yourself more about others—and have self-compassion (*see* 'Step into your shoes before you step into someone else's shoes').

If you think children with low self-image always benefit from praise, and you are looking for some nice ways of showering them with effusive praise, don't rush to Google to search 'ways to praise a child'. You are sure to find numerous posters in various colours and shapes, but don't bother downloading them. Puffed-up praise does more harm than good.

Eddie Brummelman, a psychologist at the Utrecht University in the Netherlands, and colleagues say that children are often lavished with inflated praise (for example, 'Wow, your drawing is incredibly beautiful').

Their studies involving groups of about 1000 adults and 500 children show that inflated praise, however, can have the opposite effect on self-esteem. 'Inflated praise can backfire with those kids who seem to need it the most—kids with low self-esteem,' says Brummelman.

One of the studies involved 240 children who visited a museum. After assessing their self-esteem, the children were asked to draw a famous van Gough painting, Wild Roses. After they finished their paintings, children were given a card from someone identified as a 'professional painter' with one of the three responses: 'You made an incredibly good painting' (inflated praise), 'You made a beautiful painting' (non-inflated praise) or no comment about the painting at all (no praise).

After receiving the card, the children were asked to make other pictures, but they could choose which one they would copy, either a complex drawing or a simple one. It turned out that children with low self-esteem were more likely to choose the easier pictures if they received inflated praise. On the other hand, children with high self-esteem were more likely to choose the more difficult pictures if they received inflated praise.

'If you tell a child with low self-esteem that they did incredibly well, they may think they always need to do incredibly well,' says Brummelman. 'They may worry about meeting those high standards and decide not to take on any new challenges.' Lesson in brief is fight your urge to give inflated praise to children with low self-esteem. Give them frank, straightforward praise

Time to check your Facebook status.

A study by Jeffrey Hancock of Cornell University on Facebook's psychological effects shows that Facebook boosts self-esteem. 'Unlike a mirror, which reminds us of who we really are and may have negative effect on self-esteem if that image doesn't match with our idea, Facebook can show a positive version of ourselves,' he says. 'We're not saying a deceptive version of self, but it's a positive one.'

Some may describe seflies on your Facebook pages as sign of self-indulgence or attention-seeking behaviour but they are harmless as long as they are not the only source of your self-esteem. 'Selfie' is the Oxford English Dictionary's 'Word of the Year 2013'. It's defined as a 'photograph that one has taken of oneself, typically with a smartphone or webcam and uploaded to a social media website'.

Active participation on Facebook produces a sense of belonging, but if you are being ignored, the rejection may have adverse effects. A team of University of Queensland psychologists led by Stephanie Tobin looked at how relationships between users of Facebook impact on feelings of social belonging which in turn affects outlook on life, loneliness and self-esteem. In one of the study, participants used accounts set up by researchers and were encouraged to post and comment on the posts of others. Half the participants received the feedback while the other half didn't and were effectively ostracised. When asked by researchers how they felt, both passive and shunned users experienced feelings of exclusion and felt 'invisible' and less important as individuals. Sunned users also experienced lower self-esteem. 'Being ignored makes everyone feel rotten,' says Tobin, 'it doesn't matter how sensitive you are.'

'Facebook had to be the biggest playground for self-absorbed assholes that the world had ever seen.' (Jana DeLeon, *Louisiana Longshot*). You don't have to agree, it's a novelist's opinion, not an opinion based on evidence.

Rosenberg self-esteem scale

The scale, developed in 1965 by American sociologist Morris Rosenberg, is one of the most widely used self-assessment scales. For each statement: write SA if you strongly agree; A if you agree; D if you disagree; and SD if you strongly disagree.

1. On the whole I am satisfied with myself.
2. At times, I think that I am no good at all. *
3. I feel that I have a number of good qualities.
4. I am able to do things as well as most other people.
5. I feel I do not have much to be proud of. *
6. I certainly feel useless at times.*
7. I feel that I'm a person of worth, at least the equal of others.
8. I wish I could have more respect for myself.*
9. All in all, I am inclined to feel that I am a failure.*
10. I take a positive attitude towards myself.

Scoring SA=3, A=2, D=1, SD=0. Items with an asterisk are reverse scored, that is, SA=0, A=1, D=2, SD=3. Add the total for all items. The higher the score, the higher the self-esteem.

The scale ranges from 0 to 30. Scores between 15 and 25 are within normal range; scores below 15 suggest low self-esteem and, if you wish, an opportunity to work on it (not with the help of people like Jean-Baptiste Clamence who are 'always bursting with vanity' and would work hard to push their score up and yours down).

Chapter 35

Don't eat that doughnut!

learn science of self-control/*unlearn* self-control has no downside

> 'Reason sits firm and holds the reins, and she will not let the feelings burst away and hurry her to wild chasms.'
>
> – Jane to herself when she mistakenly believes that Mr Rochester is going to marry Miss Ingram
>
> Charlotte Brontë, *Jane Eyre*

Remember Cleanthes, the Stoic (who appeared in the story 'Don't just chalk it up to circumstances')? New knowledge about the brain may have relegated many maxims of ancient philosophers to the museum of curiosities, but the Stoic assertion that willpower is like muscle power stills holds true. The more you exercise your will, stronger it becomes.

Roy Baumeister, an expert on self-control and co-author of *Willpower: Rediscovering the Greatest Human Strength*, agrees: 'People have said for centuries that you can build character by making yourself do things you don't want to do, that by exercising self-discipline you can make yourself into a stronger person. That does appear to be correct.' He also points out that willpower can wilt when your mind is under stress because it has been struggling at self-control for too long a stretch at once.

The brain has certain circuitry that enables a certain amount of self-control, which makes choice possible. However, self-control various from person to person. Says Patricia Churchland, an American neurophilosopher, or philosopher of neuroscience, and author of *Touching a Nerve: The Self as Brain*: 'Any teacher will tell you that. Some people need to struggle to achieve self-control and self-discipline. Even rats differ. Some can defer gratification and some are not so good at all.'

The Stoics say that it takes effort to exercise self-control; and not exercising self-control also takes effort. Self-control is all about battle for dominance between two front regions the brain: the region where reward and pleasure are processed (it releases dopamine which makes you feel good); and the region related to self-control (it inhibits response).

When it comes to resisting daily temptations, your brain tends to give in. Richard Lopez, a cognition scientist at Dartmouth College in the US, and colleagues recruited 31 women to take part in two brain-imaging sessions. In the first session, the participants looked at various pictures of high-calorie and fast food items and said loudly whether the pictures were taken indoors or outdoors. This session measured activity in the reward-related brain region. In the second session, the participants pressed buttons to indicate whether the pictures that flashed at intervals of just 2.5 seconds were of food items or non-food items. This task measured activity in a brain region related to self-control. For one week after the brain-imaging sessions, the participants were involved in so-called 'experience sampling' in which they used smartphones to report their food desires and eating patterns.

Participants who had relatively higher activity in the award-related region of the brain in response to the food pictures experienced more intense food desires. They were also more likely to succumb to their food craving and eat the desired food. But participants who showed relatively higher activity in brain region related to self-control acted on their cravings less often. The battle for dominance between these two brain regions—pleasure and reward region where the 'happy hour' bartender serves and self-control region where the bouncer lurks—can be fierce. When dieters exhaust their self-control, it heightens reward-related brain activity, effectively 'turning up the volume on temptations'. The researchers believe their research may ultimately help people learn ways to resist their temptations. 'Failures of self-control contribute nearly half of all death in the United States,' they note.

A sure way for dieters to boost their self-control is to follow Dutch psychologist Mirjam Tuk's advice: try to hold off from visiting the bathroom for a few hours. Her research shows that people with a full bladder make more informed decisions.

Bladder control is the first physiological determining factor that leads to an increase in self-control rather than a decrease. For her self-control, Tuk was awarded the Ig Nobel Prize for Physiology or Medicine in 2011. Ig Nobels, a satirical version of the traditional awards, are organised by the scientific journal, *Annals of Improbable Research*. They honour achievements that first make people *laugh*, and then make them *think*.

If you do have to rush to the bathroom, other studies suggest, walking backwards or tensing muscles can also improve your self-control. But don't cross your arms. If you've started eating that doughnut, you're likely to go for the second one too. A European study suggests that arm crossing leads to better performance at a task at hand.

Forget bladder control or walking backwards, think of God if you want to boost your self-control. In a study by psychologists at Queen's University in Canada, participants were tested on their self-control by asking them to endure discomfort to earn a reward or delay immediate payment to obtain a large reward. In the first test, half of the participants solved word puzzles with religious themes to prime their subconscious thoughts of religion. Then the participants were asked (using a ruse) to drink an unsavoury mix of orange juice and vinegar. The more drink they forced down, the greater their self-control. Almost twice as many of those primed with religious thoughts drank more brew and opted for money later.

The researchers suggest that the primary purpose of religious belief is to enhance the cognitive process of self-control

If you are a non-believer (or a believer): surround yourself with strong-willed friends. 'People with low self-control could relieve a lot of their self-control struggles by being with an individual who helps them,' says psychological scientist Catherine Shea of Duke University who led laboratory and field studies to determine the link between self-control and dependence on work colleagues and romantic partners. People with low self-control can get by with a little help from their friends.

If the above advice has improved your self-control and you're now on a strict diet, check your blood glucose level before you have a serious conversation with your partner. Psychologists at Ohio State University recruited 107 married couples and equipped them with blood glucose meters, voodoo dolls and 51 pins to record their glucose level and anger level. Anger level was determined by how many pins they stuck into their dolls just before going to bed when their partner wasn't looking. After 21 days the participants competed in a computer game that allowed them to blast their partners with an unpleasant noise—such as a fingernail scratching a chalkboard, ambulance sirens, dentist drill—as loud and as long as they wished. The researchers say that their findings suggest a connection between blood glucose level and self-control: glucose provides the energy the brain needs for self-control; when glucose levels are low, aggression is more likely. Some experts disagree with the results, just like partners in the study, we presume.

Self-control may be one of the most cherished values, but could too much self-control have a downside? Tufts University psychologists Evan Apfelbaum and Samuel Sommers explored the virtue of powerlessness in race relations. They thought that well-intentioned people are careful—sometimes too much careful—not to say the wrong thing about race in a mixed-race group. They also thought that too much self-control might actually cause both unease and guarded behaviour, which could in turn be seen as racial prejudice. To test this, they gave a group of white participants a series of computer-based mental exercises that were so challenging that they temporarily depleted the cognitive resources needed for self-control. They then put the participants (and others not so depleted participants) into a social situation with the potential for racial tension—they met with a black interviewer and discussed racial diversity. The participants who had been mentally fatigued were less inhibited while talking to black interviewers and their conversation was much more enjoyable than those with their self-control intact. The black interviewers saw the participants with low self-control as less prejudiced against whites.

Obviously, Jane wasn't fatigued and her self-control was intact when she mistakenly believed Mr Rochester was going to marry Miss Ingram. She continues:

'The passions may rage furiously, like true heathens, as they are; and the desires may imagine all sorts of vain things: but judgment shall still have the last word in every argument, and the casting vote in every decision.'

Charlotte Brontë, *Jane Eyre*

When reason rules passion, that's total self-control.

Chapter 36

Pop antioxidant pills with a grain of salt

learn popping pills packed with antioxidants can do more harm than good in otherwise healthy people/*unlearn* to believe too devoutly in antioxidant-rich foods

> 'Come on, it's my first day. I want to make a good impression. And clearly biology can't be understood without lipstick,' Luke joked.
> 'Funny.' Eve grabbed the lip glaze back. 'This stuff is really good for you.'
> Luke raised his eyebrows. They disappeared into his floppy blond hair.
> 'It has green tea antioxidants,' Eve continued. 'And macadamia extract and aloe vera for healing.'
>
> Amy Meredith, *Shadows*

While Eve and Luke are off to their biology class, you are welcome to my biology class. Free green tea, full of antioxidants, for all. In a cup, not in a capsule.

Highly reactive chemicals called free radicals are formed naturally in the body. Although they play a friendly role in many body processes, at high concentrations they turn into highly destructive enemy agents wrecking havoc on cell proteins and membranes including DNA. Over a lifetime, the damage to DNA may lead to the development of cancer, Alzheimer's disease and other health conditions.

Antioxidants in foods mop up rogue free radicals. According to the US Department of Agriculture, 22 fruits and vegetables with highest concentrations of antioxidants are: prunes, raisins, blueberries, blackberries, strawberries, raspberries, plums, oranges, red grapes, cherries, kiwi fruit, pink grapefruit, kale, spinach, brussels sprouts, alfalfa sprouts, broccoli flowers, beets, red bell pepper, onion, corn and eggplant. Other good sources of antioxidants include: green tea, dark chocolate, red wine, vegetable oils, whole grains, avocados, pomegranate juice, cloves, oregano, thyme, turmeric, walnuts and black rice. All these foods should be part of any balanced diet, so it makes sense to pack antioxidants into pills. Popping these pills is sure to keep us healthy. Or would it?

The story of free radicals began in 1945 when wife of Denham Harman, then a chemist working at Shell Oil in the US, showed him an article 'Tomorrow you may be younger' in *The Ladies' Home Journal* which sparked his interest in ageing. Ten years later he published his first paper linking ageing to free radicals. At that time free radicals were almost unheard of in medicine; doctors worried more about germs. Other studies showed that diets rich in food-based antioxidants muted the damaging effects of free radicals.

Food supplement and vitamin manufacturers were quick to join the dots—taking antioxidants are good for your health—and started filling grocery shelves with pills loaded with considerable doses of antioxidants such as vitamin E and beta-carotene (a red-orange pigment found in plants and fruits which the body converts into vitamin A, an antioxidant). Why bother eating your veggies? Shut up, Mum! Antioxidant has now become a household buzzword, even finding its way into teenage novels.

Sip your green tea and get ready to bust some antioxidant myths:

Free radicals are always bad. Research now shows that in certain amounts and situations, free radicals might not be dangerous but useful and healthy. When it comes to diabetes, free radicals can be good for you. Antioxidants might make the things worse because some of them block the beneficial effects of exercise.

Antioxidants are always good. Many clinical trials have concluded that antioxidant supplements do not reduce risk of death. Many health organisations now advise that people should not take antioxidant supplements except to treat a vitamin deficiency that has been diagnosed by a doctor. 'Although antioxidant supplements are not recommended, antioxidant food sources—especially plant-derived foods such as fruits, vegetables, whole-grain foods and vegetable oils—are recommended,' says the American Health Foundation.

Antioxidants are all vitamins. There are thousands of antioxidants, but only some vitamins such as A, C and E are antioxidants.

All antioxidants are the same. Each antioxidant is unique and fights only a certain free radicals. For example, vitamin C or ascorbic acid is soluble in water and attacks only those free radicals that are in an aqueous environment such as inside cells.

If antioxidants are good, more are better. Studies show that too much antioxidants in diet can become pro-oxidants, meaning they create more harmful free radicals.

The brain is highly susceptible to damage caused by free radicals. It consumes a lot of energy (about 20 per cent of our daily calorie intake) and the reactions that release this energy also produce free radicals. But antioxidants may not really be the brain foods as a new study suggests that not all diets high in antioxidants reduce the risks of dementia and stroke.

The study involved 5395 people age 55 and older who had no signs of dementia at the start of the study. Researcher

Elizabeth Devore of Harvard Medical School and her colleagues determined each participant's antioxidant score based on a questionnaire about how they ate 170 foods over the past year at the start of the study. Then the participants were followed for an average of nearly 14 years. About 600 people developed dementia during the study and a similar number had a stroke. But researchers found that people with high levels of antioxidants were no more or less likely to develop dementia or suffer a stroke than with people with low levels of dementia. The participants were from the Netherlands and the variability of their antioxidant levels was due to the amount of tea and coffee they drank. Coffee and tea contains flavonoids, a group of compounds that contain antioxidants.

However, a similar study on a group of older Italians found that higher antioxidant levels were linked to a lower risk of stroke. The antioxidants in the diet of Italians were mainly from eating fruits and vegetables. 'It is not about total antioxidant levels,' says Devore, 'it is about specific antioxidant-rich food.' There are a lot of studies that suggest that higher fruit and vegetable intake is linked with lower risk of stroke. Your mother was right when she coerced you to eat more veggies.

In a prostate cancer trial men who took vitamin E for more than five years had a 17 per cent greater risk of developing the disease than men who took a placebo. Research on mice has found an explanation, at least for lung cancer: antioxidants help early tumours survive and grow by protecting them and their DNA from damage from free radicals. Antioxidants wee not only protecting health cells, they were also protecting cancer cells from bodies' defences. The clear message from cancer researchers is: stick to healthy balanced diet without taking supplements.

'I believe the question must be asked as to whether daily consumption of antioxidant foods and pills significantly heightens the risk not only of cancer but also type 2 diabetes,' says James Watson, who in 1953 co-discovered with Francis Crick the structure of DNA. Watson agrees that testing on animals is unlikely to provide results of sufficient power or persuasiveness to settle the question.

Cochrane Reviews are the most respected and impartial evaluations of medical research. In 2012 a systematic review of 78 clinical trials involving 296,700 participants found no evidence to support antioxidant supplements (beta-carotene, vitamins A, C and E and selenium) for primary or secondary prevention. Instead the review showed higher doses of beta-carotene and vitamins A and E seem to increase mortality. 'Antioxidant supplements need to be considered as medicinal products and should undergo sufficient evaluation before marketing,' the review advised.

The question of the value of antioxidants has not yet been settled finally but message of my 'biology lesson' is easy to digest. When we eat, say broccoli, our body absorbs its vitamin C and beta-carotene and disperses them in a natural way. If the same vitamin C and beta-carotene are concentrated in a capsule, their dispersal is not normal and may even lead to generation and build up of harmful free radicals. What's better, a well-balanced diet rich in vitamins and antioxidants or consummation of junk food supplemented with capsules?

Remember preventing a disease is almost curing it (without any expensive treatment). Forget Eve. This stuff is *not* really good for you. A balanced diet and regular exercise are still good for you.

Chapter 37

a bottleful of empty promises

learn to rethink the need for vitamin supplements/*unlearn* vitamins are always good for you

> Jocelyn goes, Watch, Rhea. They'll be blond, her sisters.
> I go, According to?
> Rich children are always blond, Jocelyn goes. It has to do with vitamins.
>
> Jennifer Egan, *A Visit from the Goon Squad*

Author goes, children are not necessarily blond (or any other hair colour you fancy), even if they are born in rich families and pop multivitamin pills daily. There is no evidence either that multivitamins make children smarter. Forget about multivitamins and sit down daily with your toddler and a good book for some interactive reading. It's a sure way of making them smarter rather than forcing sugary multivitamin syrup down their throats, unless advised by a doctor.

Zillions of healthy adults also take multivitamin and minerals supplements apparently believing blindly such supplements are essential for good health. A multibillion-dollar industry is built on this myth. Our bodies need vitamins to grow, develop and function normally. Vitamins are natural substances and most foods contain a little or more of some of the vitamins. A balanced diet usually provides all of the vitamins required. However, there are times, such as during pregnancy, childhood or certain illnesses when health professionals prescribe additional vitamins.

Our obsession with vitamins can be traced back to Linus Pauling, the only person to have been awarded two unshared Nobel Prizes—the chemistry Nobel in 1954 and the peace Nobel in 1962. In 1970 Pauling, who took up to18,000 milligrams of vitamin C daily, published a book, *Vitamin C and the Common Cold*, which became an instant bestseller. In this book he urged people to take 3,000 milligram of the vitamin daily (the current recommended dietary allowance or RDA is only 60 milligrams; a large glass of orange juice contains about 100 milligrams). This highly popular and influential book's recommendation was not based on much scientific evidence.

The common cold is not a precisely defined disease, yet most of us are familiar with its symptoms and believe that vitamin C can kill the cold. A 2013 Cochrane Review, the most respected and impartial evaluations of medical research, affirms that large doses of vitamin C have little preventive effect and are not effective in dealing with the symptoms. The review analysed 29 clinical trials carried out around the world over six decades and involving 11,300 patients who took 200 milligrams or more of vitamin C daily. 'Routine vitamin C supplementation is not justified, yet vitamin C may be useful for people exposed to brief periods of severe physical exercise,' the review advised.

Two decades after publication of his book Pauling made a new claim that vitamin C, taken with massive doses of vitamin A (25,000 international units or IU, which is nearly 100 times more than the RDA), vitamin E (400 to 1,600 IU – the RDA is only about 20 IU), beta-carotene (the body converts it into vitamin A) and selenium (basic element) could not only cure colds but cancer as well. The prestigious *Time* magazine lent a helping hand in perpetuating this falsehood when its cover story of 6 April 1992 pronounced, 'The Mighty Vitamin: More may be better when it comes to fighting cancer, heart disease and aging'.

The mighty vitamin was unknown until 1906 when English biochemist Frederick Hopkins noticed that his laboratory rats failed to grow when fed on a diet of pure fats, proteins and carbohydrates, but they grew rapidly when he added even a tiny amount of yeast extract to their diet. He suggested that diseases such as rickets and scurvy were caused by this missing dietary substance. Hopkins continued his research and later identified two mysterious components which became known as vitamins A and D.

Interestingly, when in 1932 Hungarian biochemist Albert Szent-Györgyi submitted his paper announcing the discovery of vitamin C to the journal *Nature* he named it 'ignose' as it was a sugar (sucrose, fructose, glucose etc.) of unknown composition. The editor considered the term too flippant and rejected it. Szent-Györgyi sent back his paper with the compound renamed 'godnose'. Only God knows why so many people around the world are fascinated with vitamin C and its 12 'apostles' in the human diet: A, the eight B vitamins (originally thought to be one vitamin), C, D, E and K. They all are essential but only in the right quantities.

The US Preventive Services Tasks Force, a panel of independent experts in prevention and evidence-based medicines, claims that in the US about half of the adults have used at least one dietary supplement and one-third have used a multivitamins. More women than men use supplements, and more older adults than younger adults use them. People take these supplements to improve or maintain overall health. In 2014, the Task Force concluded 'there is not enough evidence to determine the effectiveness of taking vitamins and minerals to prevent cardiovascular disease and cancer.' It also concluded 'the evidence shows that there is no benefit to taking vitamin E and that beta-carotene can be harmful because it increases the risk of lung cancer in people who are already at risk to the disease'. It also repeated the well-known mantra: 'Adequate nutrition by eating a diet rich in fruits, vegetables, whole grains, fat-free and low-fat dairy products, and seafood have been associated with a reduced risk of cardiovascular disease and cancer.'

A study by a team of 12 cancer experts in the US led by Alan Kristal of Fred Hutchinson Center in Seattle has found 17 per cent increased risk of prostate cancer in healthy men from vitamin E supplements. Selenium supplements appeared to have no effect.

A popular urban myth is that if some vitamins are good, then more must be better. However excessive intake of some vitamins can be toxic for the brain. Doctors call it hypervitaminosis, which is usually caused by fat-soluble vitamins A, D, E and K. Vitamin Bs are water soluble.

Studies show two other risks of taking vitamins supplements. First, those who take vitamin supplements may be more likely to take risks with their health; pill-poppers tend to exercise less and eat more junk food. Second, vitamin supplements seem to give people license to indulge; they reward themselves with an unhealthy treat after taking supplements.

In spite of so much popping of vitamin pills, estimates show that 20 to 80 per cent of the population worldwide have insufficient vitamin D intake. This vitamin has long been known as important for bones, but new studies show low levels of vitamin D may impair mental abilities. Healthy adults need 1,000 to 2,000 IU daily—our bodies can synthesise this amount from 30 to 40 minutes of sum exposure two to three times a week. Some skin types absorb less vitamin D and it may be worthwhile for older people, especially women, to see their doctor whether they need extra vitamin D. Taking vitamin D supplements without your doctor's advice is not only a waste of money, excessive intake may even harm you. Strict vegetarians also need vitamin B12 supplements. Taking vitamin B6 (folic acid or foliate) supplements during frequency is also recommended, but seek your doctor's advice for the amount of daily dose.

Carlos Ruiz Zafón writes in *The Angel's Game*:

> Don Basilio was a forbidding-looking man with a bushy mustache who did not suffer fools and who subscribed to the theory that the liberal use of adverbs and adjectives was the mark of a pervert or someone with a vitamin deficiency.

More appropriately, Don Basilio should have subscribed to the theory that the liberal use of adverbs and adjectives was the mark of someone who pops multivitamin pills daily.

Chapter 38

We certainly are what we eat

learn what you eat controls your body and mind/*unlearn* misconceptions about food

> Why are bodies so difficult to manage? Why? 'Oh, oh, look at me, I'm a body, I'm going to splurge fat unless you, like, STARVE yourself and go to undignified TORTURE CENTRES and don't eat anything nice or get drunk.' Hate diet.
>
> Helen Fielding, *Bridget Jones: Mad About the Boy*

Yes, Bridget Jones, our bodies are difficult to manage because we embrace foods that are at odds with our own health. Our food habits are hard to change because what we like began the very day we were born. The result of a classic experiment is often cited as a proof that our food habits are formed even before we are born: babies are more likely to show a preference for carrot-flavoured cereal if their mother drank carrot juice during pregnancy or while breastfeeding. Biology can't be changed, but there is enough inherent plasticity in our bodies and brains to change to make us like healthy foods. What follows is a mini survey of latest findings to veer you away from Bridget Jones' 'Hate diet' idea.

Count hormones, not calories

Food we eat is made up of complex molecules. Our digestive system turns them into simpler molecules such as glucose and amino acids and releases them into the blood stream. Once in the blood stream glucose can be converted immediately into energy or stored in our bodies as fat to be used later. We measure the available energy in all foods in calories. Fats provide about nine calories per gram, alcohol seven, carbohydrates and proteins four, and fibre delivers just two calories.

This calorie count is based on 19th-century experiments in which food was placed in a sealed container surrounded by water. It was completely burned and the resulting rise in temperature was measured to calculate food calories. The system in use today is based on indirect calorie estimation calculated by adding up the calories provided by energy-containing nutrients: protein, carbohydrate, fat and alcohol. The system assumes that the calorie values accurately reflect how much energy different bodies derive from different kinds of food.

New research shows that this system is too simplistic. 'To accurately calculate the total calories that someone gets out of a given food,' says Rob Dunn, a biologist at North Carolina State University, 'you would have to take into account a dizzying array of factors, including whether that food has evolved to survive digestion; how boiling, baking, microwaving or flambéing a food changes its structure and chemistry; how much energy the body expends to break down different kinds of food; and the extent to which the billions of bacteria in the gut aid human digestion and, conversely, steal some calories for themselves.' Calories you count in your diet do not consider any of these factors and it seems that the demands of calorie-controlled diet is stacked against you.

The conventional wisdom is that our body weight depends on the balance between calories consumed and calories burned; and the only way to lose weight is to consume fewer calories and expend more calories. This 'calorie in, calorie out' idea now faces challenges from scientists who think of food as a cocktail of hormones. Hormones are body's chemicals messengers, pumped out by specialist cells. They travel through blood to deliver their message to specific cells to produce other chemical or action. One of the first hormones discovered was insulin; it pulls glucose out of the bloodstream.

Food is just a pile of chemicals, which can elicit specific reactions from cells, making them much like hormones. Karen Ryan and Randy Seeley, obesity specialists at the University of Cincinnati, say that food's effects on the body are so complex and specific that a meal is almost like a cocktail of hormones.

This new thinking 'suggests that the argument over whether fat or sugar is to blame for the increasing incidence of obesity may be misguided,' Ryan and Seeley say. If they are right you can throw out all your diet books.

Not all calories are created equal

In the meantime, we still have to pay homage to queen calorie. A calorie is a calorie, this popular mantra from dieticians warning dieters to watch their calorie intake is not worth chanting as a new study suggests. After weight loss, the rate at which dieters burn calories decreases, reflecting slower metabolism. This lower rate of burning calories adds to the difficulty of weight maintenance and explains why dieters tend to regain lost weight. We can blame evolution for this mechanism. Our bodies' ability to store energy in the form of fat for future use was an absolute necessity for survival when food was scarce or unpredictable. The brain-hunger system that evolved over millions of years even erred on the side of gluttony by furnishing our brains with a system that made eating a highly pleasurable activity.

Cara Ebbeling and David Ludwig, obesity experts at Boston Children's Hospital, say that a diet with a low glycemic index (GI) is more effective than conventional approaches to burning calories at a higher rate after weight loss. GI, a scale from 0 to 100, is a measure of how quickly a carbohydrate is digested and released into the bloodstream as glucose. Low-GI carbohydrates digest slowly, helping to keep blood sugar and hormones stables. Using the state-of-the art technology to measure energy expended by participants as they followed low-fat, low-GI and low-carbohydrate diets, the researchers discovered that 'total calorie burned plummeted by 300 calories on the low-fat diet compared to low-carbohydrate diet'.

Contrary to nutritional dogma, they say, all calories are not created equal. A tweet for dieters to retweet: Low-GI diets are easier to stick, compared to low-carb and low-fat diets, which eliminate entire classes of food making them hard to follow.

The ice diet conceived by New Jersey gastroenterologist Brian Weiner is also based on the idea that all calories are not created equal. One day when he noticed that the cups of his favourite Italian ice cream listed their calorie content as 100 (25 grams of carbohydrates multiplied by four calories per gram), he suddenly realised that this calculation was wrong. What about the energy required to melt the ice? He worked out that the net energy of ice cream was only 72 calories after deducting the calories the body burns to produce thermal energy that melts the ice.

He suggests that the ingestion of one litre of ice—not ice cream loaded with sugar—per day is generally safe for otherwise heathy people. It would burn 160 calories, the amount of energy used in running one mile (1.6 kilometres). Don't eat ice while you're cooling off from exercise, advises the good doctor, as heat generated by the exercise would be neutralised by ice. Why bother running when lounging on the couch and slurping a slurpee would do the job?

Food on the brain

Scientists now agree that obesity is not a behaviour disorder. It is not caused by lack of willpower and self-control. It's not even caused by hormone imbalance. How does the brain control our eating behaviour? The system that regulates our eating behaviour is highly complex. It integrates information about the body's energy needs and the status of its fat stores, and then initiatives changes in behaviour and energy processing in response. Specialised regions of brain stimulate feelings of hunger or satiety.

According to *Scientific American* magazine, the brain receives three types of information: (1) stored energy status: leptin, a hormone secreted into the bloodstream by fat cells, indicating how much fat they contain; (2) metabolic status: circulating glucose represents energy immediately available to cells; and (3) neural and chemical signals from the gut indicating whether digestive organs are full of food. The brain then responds by changing body's food intake through appetite and satiety signals. Ghrelin, a hormone produced by glands in stomach, signals the stomach's readiness for a meal to the brain. It's often called the hunger hormone. Leptin is called the satiety hormone as it suppresses appetite.

'We have evolved an efficient brain system to help maintain a healthy and consistent body weight by signalling when it is time to eat and when it is time to stop,' says Paul Kenny, an American researcher who specialises in investigating the mechanisms of drug addiction and obesity. 'But highly appetising foods can often override these signals and drive weight gain.' Food loaded with fat and sugar sparks the releases of endorphins.

This feel-good hormone can trigger binge eating. Like addictive drugs, overeating high-calorie foods can also promote release of dopamine, which can influence decision-making circuitry of the brain. In some people, the actions of endorphins, dopamine and other hormones overcome hormonal signals to stop eating. Kenny worries that even if scientists prove that obesity results from an addiction to food, 'Western societies, saturated in fat and temptation, will make it hard for any obese person to quit'. Lovers of cheesecakes take note: overweight rats in Kenny's laboratory found it hard to resist temptation of not eating cheesecake, which contained high quantities of both fat and sugar.

Good news for those addicted to their cups of caffeine. Caffeine addiction is not like food addiction. Caffeine is very similar to adenosine, a molecule naturally present in the brain. Adenosine produced over time produces a feeling of tiredness. Caffeine locks into the same brain cell receptors used by adenosine. By blocking these receptors, caffeine generates a sense of alertness and energy for a few hours. Additionally, the brain's own natural stimulant dopamine also works more efficiently when adenosine receptors are blocked.

The brain chemistry of people who regularly drink coffee, tea or other caffeine drinks changes over time. This explains why these people become jittery when they cannot get their regular fix. Good news is that, unlike many drug addictions, these change are relatively short term. To kick the habit you only need to go through seven to 10 days of restlessness without drinking any caffeine.

Food for thought

Should you cut down the amount of gluten in your diet? The answer is an unequivocal 'no' unless you have coeliac disease, non-coeliac gluten sensitivity or wheat allergy (wheat allergy can be caused by wheat constituents other than gluten). Gluten is a composite of two proteins, glutenin and gliadin, found in many grains, including wheat, barley and rye. A gluten-free diet could even be unhealthy for otherwise healthy people. Gluten-free products are often made with refined grains and are low in fibre, iron, folic acid, niacin, thiamine, riboflavin, calcium, vitamin B12, phosphorus and zinc. Also, they are more costly and some are higher in salt and fat.

Red wine and dark chocolate are touted as health foods, but a major study shows that the antioxidant resveratrol, found in red wine, dark chocolate and berries, has no significant effect on lifespan, heart disease of cancer. These foods may still be good for you, but resveratrol is not the reason.

Despite the negative results, the consumption of red wines, dark chocolate and berries appears to protect the heart in some people. 'It's just that the benefits, if they are there, must come from some other polyphenols or substances found in those foodstuffs,' says Richard Semba, an ophthalmology professor at Johns Hopkins School of Medicine who led the study. His international team studied the effects of ageing in a group of people in the Chianti region of Italy.

You may open that classic bottle of Chianti in a straw basket and say 'Salute!'. But throw away the bottle of resveratrol supplements. No study has found any benefits associated with these supplements.

Memorable foods

The more chocolate a population consumes the more Nobel Prize winners it has. This chocolate-coated conclusion was made by New York cardiologist Franz Messerli. He ranked a list of 22 countries in terms of Nobel Laureates per 10 million people and comparing it to the information on per capita chocolate consumption. The results showed a 'surprisingly powerful correlation' between the amount of chocolate consumed in each country and the number of Nobel Laureates it produced, says Messerli. Take the finding with a grain of salt, or with a peanut or two.

Cocoa contains flavonoids, a group of plant-based compounds that some studies have linked to memory, learning, reasoning skills, decision-making, verbal comprehension and numerical ability. They have also shown to slow age-related decline in mental function. Besides cocoa, flavonoid-rich foods include berries, grapes, citrus fruits, spinach, peppers, onions, celery, sage, parsley, sage, rosemary, thyme, red wine and soy foods such as tofu.

It's time to recoup the calories you have burned while reading this story. Follow Ignatius J. Reilly, the obese hero of John Kennedy Toole's comic masterpiece, *A Confederacy of Dunces*: 'When my brain begins to reel from my literary labors, I make an occasional cheese dip.'

Chapter 39

Concerns about faith

learn, whether you are a believer or nonbeliever, religious belief is part of human nature/*unlearn* belief is opposite of reason; and if you are a believer unlearn to be intolerant of nonbelievers

> 'Atheism Is Also A Religious Position,' Dorfl rumbled.
> 'No it's not!' said Constable Visit. 'Atheism is a denial of a god.'
> 'Therefore It Is A Religious Position,' said Dorfl. 'Indeed, A *True* Atheist Thinks Of The Gods Constantly, Albeit In Terms of Denial. Therefore, Atheism Is A Form Of Belief. If The Atheist Truly Did Not Believe, He Or She Would Not Bother To Deny.'
>
> Terry Pratchett, *Feet of Clay*

If you would have come along with me for a walk from my home to my college in the narrow, steep streets of my hometown in the foothills of the Himalayas—decades ago—you would notice that for about half a mile my head is bowed and I walk solemnly. This 'redemption walk', as I call it, starts when I come near the bend with the large banyan tree. I always see a few cows lazing under the sacred tree, enjoying the freedom and security our nation bestows upon them, more than it does on its many of disadvantaged citizens. When one of the cows decides to go for a walk in the middle of the street, it turns itself into a floating traffic island.

As I go around one such traffic island, come in view the graceful curves of arabesque motifs on the dome of a majestic mosque, a marvel of Mughal architecture in our corner of the world. Through the arched portal I see a vast expanse of chequered marble floor on which people pray. At prayer time the mosque is crowded with men in white skull caps kneeling with their faces towards Mecca.

After a few minutes' walk the street becomes quieter and steeper and I'm now in front of a Catholic church. Through the large stained-glass windows I see Christ on the cross, exquisitely carved in white marble. The locals call it the Begum's church: it was built by a Muslim princess more than a century ago. I never took interest in finding out why. I suppose it was a Taj Mahal built for her Christian husband—or lover.

Now I turn left into a wider street and walk downhill past a gurdwara, a Sikh temple, a simple whitewashed building proclaiming that elegance is ingeniously simple. From the street I see Guru Granth Sahib, the scared scriptures, resting on a low table covered with a white silk cloth. An old man with a long snow-white beard and blue turban always sits cross-legged in front of the table. In his right hand is a white fluffy fan that he waves gently over the holy book.

About hundred yards further on I see an ornate Hindu temple, full of statues dressed in garish clothes. Sweet smells of burning incense and enchanting sounds of prayers remind me of my karma in this world. Not far from the temple is a bustling bazaar of second-hand shops selling everything from car parts to books. My 'redemption walk' ends when I reach the bookshops. I put on my sunglasses and my demeanour changes. I turn into my usual self.

Why does an agnostic like me show such respect for these places of worship? The answer is imprinted on my left cheek. One day, when I was nine, I went for a walk with my grandfather, a not-so-devout-Hindu. As we walked past the mosque, he folded his hands and bowed his head. His stern eyes demanded the same from me. I obliged. He did the same in front of the church and the gurdwara. I reluctantly followed him. When we reached the temple, I exploded, 'Why we have to bow in front of every stupid temple?' I ran inside the temple and spat on a statue of a god.

My grandfather grabbed me and smacked me on my left cheek with a force I thought only Hercules was capable of applying. 'I don't care whether you believe in any religion, but you must always respect all religions,' the words erupted from his mouth like lava from a volcano. My grandfather is a man of peace and non-violence, the Mahatma Gandhi of our family. He always exudes happiness and contentment, and I have never seen him angry.

After so many decades, burning sensation in my left cheek is now a fading memory, but reverberations of 'respect all religions' still echo in my ears whenever I walk past a place of worship. Yet faith eludes me even when I imagine my karmic fate after my death according to the religion of my ancestors:

> Yama, the god of death, leads me to Mount Kailasha, the abode of Lord Shiva and his wife Parvati, daughter of the mountain. Nandi, the bull who watches over their gate, lets me go in. As I enter the gate Ganesha, the four-armed, elephant-headed god of wisdom, the son of Shiva and Parvati, directs Nandi to take me first to Dharma Rai, the divine accountant. Dharma Rai takes into account our past deeds, makes sure that we have paid our karmic debts and accordingly decides when, where and how we have to be born again. 'As you have not practised your religion,' he says. 'I deny you nirvana. You'll be born again; but because of your good karma in the past life, in a higher form of life than a human.' 'I would be born as a Bollywood demigod then, Dharma Rai,' I cry with joy. 'No,' says Dharma Rai in his beancounter's flat voice, 'you would be reborn on the planet of the apes.' 'As Hollywood god Charlton Heston on Planet of the Apes?' I ask anxiously. 'I'm sorry,' replies Dharma Rai with a slight smile, 'your karma gives you only an extra's role as a chimpanzee.'

The rise in the numbers of nonbelievers—some estimates show their numbers to be nearly one billion worldwide; a number surpassed only by adherents of Christianity and Islam—is driven by many factors; most notably, by a sense that religion is out of step with changing values. It would seem that reason is winning over faith, if we subscribe to the view that faith and reason are in conflict. They are not. It's possible to strike a balance between faith and reason. 'Faith is not the opposite of reason,' observes Francis Collins, a leading genetic scientist and a religious believer. 'Faith rests squarely upon reason, but with the added component of revelation.' Einstein's famous aphorism 'science without religion is lame, religion without science is blind' also reconciles faith and reason. Einstein, however, could not accept a deity who could meddle at whim in the events of his creation or in the lives of his creatures.

These days big machines enable neuroscientists to peer into our brains. Is there a difference between the brain of a believer and the brain of a nonbeliever? The answer is a qualified 'yes'.

Brain scans of long-term meditators show changes in the actual structure of the cerebral cortex, the outer parts of the brain usually referred to as the grey matter or thinking cap. Associated with reasoning and sensory perception, these parts are found to be thicker in meditators than non-meditators. The thickening was more pronounced in older than in younger people. This has intrigued researchers because these parts of the cerebral cortex normally shrink as we age.

Those deeply involved in meditation show the greatest thickening, confirming that it was caused by extensive practice. People who meditate not only show more grey matter, they also have stronger connections between different regions. These connections increase the ability of neurons to rapidly relay signals in the brain. Studies show the increase to be throughout the brain, not just in certain areas.

Studies also show that people who have had 'born again' experiences have a smaller hippocampus than nonbelievers do. The hippocampus, a region of the brain, which is a continuation of the cerebral cortex, is known to be involved in emotions and memory.

There is nothing unique about religious belief in the brain. But in a believer's brain, God is as real as any object or person. Experiments show that thinking about God people of different beliefs all tend to use the same regions of the brain, which they use when they think about something else. There is nothing unique about praying to God either; it's just another kind of friendly conversion. Whether you are reciting a prayer or nursery rhyme, the same brain areas, typically associated with rehearsal and repetition, are activated.

'For now we can say that the religious and atheist brains exhibit differences, but what causes these disparities remains unknown,' concludes Andrew Newberg after studying about 150 brain scans, including of those Buddhists, nuns, atheists. Newberg is an American neuroscientist and author of many books, including the ground-breaking, *The Metaphysical Mind: Probing the Biology of Philosophical Mind.* Religion doesn't make you happy supernaturally. Your beliefs, whether you're a believer or a nonbeliever, make your brain release dopamine, a chemical that controls the brain's sense of reward and pleasure. That's the key to happiness.

After assessing stress levels and degree of belief in religion and science of 54 rowers, a team of psychologists led by Miguel Farias at the University of Oxford also agree that any kind of belief system helps you to structure your perception of reality. 'It allows you think of the universe in a particular meaningful way,' says Farias.

Would belief in Einstein help me during moments of stress? I don't know. But he has helped me to understand the universe—the physical, not the spiritual—in a meaningful way. But again, religion alone doesn't make you happy as it doesn't affect happiness supernaturally. 'It has to happen through psychological, sociological and biological mechanism,' says psychiatrist Harold Koenig of Duke University in the United States.

Vassilis Saroglou, professor of psychology at the Université Catholique of Louvain in Belgium, who has studied the influences of genes and personality on our attitudes towards religion, believes that we are born to be inclined towards religion or atheism: 'Does God call us? For some of us, the answer is yes: through our genes, parents, acquaintances and life events.'

He seems to provide an answer to my dilemma when he says: 'The genetic influences help to explain why adults sometimes stray from the beliefs of their childhood. The more distance they get from the influences of their early years, the more idiosyncratic factors can hold sway over a person's attitudes.'

Prayer is an integral part of religion. Nonbelievers say that with countless prayers having been said for thousands of years, we should expect some miraculous effects of prayers. There are no verifiable answers from God. C. S. Lewis (*Chronicles of Narina*) provides an explanation: 'For prayer is request. The essence of request, as distinct from compulsion, is that it may or may not be granted.'

'You know, you're are either a person of faith, or not,' says Reza Aslan, author of *Zealot: The Life and Times of Jesus of Nazareth*, 'You either believe that there's something beyond the material world that you can commune with, or you don't. If you *do* believe it, then it helps to have a language to help you express that ineffable experience—to yourself, and to other people. And that is all that religion is—that language.'

I may not have that divine language, but I at least I do have an earthly language that let me express equal respect for all religions, thanks to the memory of burning sensation in my left cheek. In *Feet of Clay*, who is right? Dorfl or Constable Visit? You decide.

Chapter 40

a sure thing

learn not to fear death; death is inevitable/*unlearn* to be afraid of what it would like to be dead; there is no afterlife

> 'So tonight to shush you how about it if I say I have administrative bones to pick with God, Boo. I'll say God seems to have a kind of laid-back management style I'm not crazy about. I'm pretty much anti-death. God looks by all accounts pro-death. I'm not seeing how we can get together on this issue, he and I, Boo.'
>
> – Hal Incandenza to Mario ('Boo') Incandenza
> David Foster Wallace, *Infinite Jest*

Advances in medicine and sanitation have helped increase average life span of humans globally for more than a century: in 1900 the world average life expectancy was 30 years; in 1950 it increased to 50 years; in 2010 to 70 years (80 years in developed countries); it's expected to rise to 75 years in 2050 (85 years in developed countries).

Why do we age? The answer to this question lies in the genes. Choosing the right ancestors would be a big help. They should be rich (wealth brings health) and should have right genes (not the genes with predisposition to disease such as cancer or Alzheimer's).

Even if we have the right genes, they cannot help us to live forever. Cells are damaged all the time, sometimes they began to replicate uncontrollably and become cancerous. This damage is accumulated over time leading to the breakdown of healthy functions of the body. To help us live longer and healthier, scientists are trying to find ways to manipulate damaged cells. An anti-ageing pill is not yet on science's horizon, but it hasn't stopped bogus health gurus and snake oil salesmen to peddle anti-ageing 'therapies'. Scientists warn that no anti-ageing remedy on the market has been proven effective.

Even the conventional wisdom among scientists and the general public that a low-calorie diet—restricting calories intake 10 to 40 per cent below normal consumption levels—can slow down ageing has been challenged. No, don't rush to your local McDonald. Regular exercise and maintaining a healthy weight is essential if you want good health up to the time of death.

Demographers estimate that about 107.7 billion people have been born since 8000 BC. We're the lucky among the seven billion people that are still alive. One day, when our heart stops beating and there is no blood flowing to our brain, all brain activity will cease and we'll be declared clinically dead. Brain death is arguably the most accurate biological and philosophical representation of death. Heart failure in itself does not constitute death. Pumping of blood is important, but it only supports the functions of the brain. Simultaneous monitoring of heart rate and brainwaves shows that it takes brainwaves 11 to 20 seconds to go flat when the heart stops pumping. Brain death, or 'irreversible coma' as it's known in the medical community, means the cerebral cortex is destroyed forever. This largest part of the brain controls everything that makes us human: sensory analysis, spatial location, language, memory, attention, emotion, motivation, thought and consciousness.

Is death instantaneous? No, doctors say that we take from a few seconds to a few minutes to go through the dying process. 'The last breath is taken, death takes hold and life is over,' says Thomas Kirkwood, author of award-winning *Time of Our Lives: The Science of Human Aging*. 'At this moment, most of the body's cells are still active. Unaware of what has happened they just carry on ... In a short while, starved of oxygen, the cells will die.'

When the cells are dead the tissues start decomposing and the body eventually reverts back to dust and only skeleton and teeth are left. Depending on environmental conditions such as temperature and humidity, it may be as quickly as two weeks or as slowly as two years.

What follows death? Do we experience our own non-existence? Do we suffer from being dead? Is there life after death? Can we know anything about afterlife? Do the souls of dead continue to exist in an afterlife world? Science is not concerned about these questions. They belong to philosophers and theologians.

The fourth-century BC Greek philosopher Epicurus, who believed that sensory enjoyment was the most important thing in life, was the first to argue that we must not fear death because we won't be able to experience our own non-existence. His followers were called Epicureans and they lived in a garden. The story goes that an inscription on the gate to the garden said: 'Stranger, here you will do well to tarry; here our highest good is pleasure.' Yet the life of community was simple and frugal. Their food and drink was mainly bread and water. Epicureans continued to flourish for another five centuries and exerted a considerable influence over all ancient philosophy and science. Ironically, epicurean now denotes someone with luxurious tastes or habits, especially in eating or drinking.

Epicurus said that death is the primary cause of anxiety among human beings. He argued that death involves neither pain nor pleasure. The only thing that we should fear is pain. Therefore, we should not fear death. 'Death, the most awful of evils, is nothing to us, seeing that, when we are dead, death is not come, and when death is come, we are not,' he said. In other words, death is nothing to us, since while we exist, our death is not, and when our death occurs, we do not exist. Shelly Kagan, a professor of philosophy at Yale University and author of *Death*, echoes Epicurus when she says fear is one of the most common reactions to death. 'Indeed, "fear" may be too weak a term: terror is more like it,' she adds. It's reasonable to be afraid of the process of dying, but most people are terrified of death itself.

On the matter of afterlife, Peter Atkins, presents a scientific answer in his excellent book, *On Being: A Scientist's Exploration of the Great Questions of Existence*: 'Quite frankly, it is hard to credit that the extraordinary property of complexly interacting organized matter we call consciousness can survive the decomposition of that matter. Science, especially through psychology, shines its brilliant light on the afterlife and instead of illuminating it causes it to shrivel and die, revealing its core: anxiety.'

Death is the final frontier and we have simply ceased to exist. People who have returned from the brink of death do not agree. They tell amazingly similar stories about their experiences. The first stage of their 'life after life' is the feeling of immense peacefulness and the absence of pain and fear. They then somehow leave their physical body and find themselves looking down upon it. They continue to rise above their body and enter into a dark tunnel. Their peaceful journey ends when they see a light at the end of the tunnel. The distant, golden light is welcoming and some regard it as a supernatural presence of some sort. They think they have reached the boundary between life and afterlife. Some even recall speaking to their dead relatives or an encounter with certain aspects of their lives. What follows is the realisation that they have to leave this afterlife and then they wake up.

These people often see their near-death experiences as paranormal experiences. There is a soul or psyche dwelling in the physical body. When we die this immaterial essence leaves the body and travels to another world. Naturally, scientists disagree. They say that the brain surges with activity just before death. The retrospective analysis of brain activity of critically ill patients as they were removed from life support has shown that there is a significant spike in brain activity at or near the time of death. This increase in brain activity, which lasts from 30 to 180 seconds, could explain near-death experiences. Some researchers associate near-death experience with the temporal lobes, the part of the brain that lies around ears. Studies suggest that electrical activity in the right temporal lobe is involved in mystical and religious experiences. Investigations of temporal lobe activity in people who had near-death experiences during life-threatening events have revealed that such a people have more temporal lobe activity than normal people.

Everyone agrees that near-death experiences have some basis in normal brain function gone off course. When Grim Reaper calls brain function goes off completely.

The foregoing is unlikely to help Hal Incandenza, the protagonist of *Infinite Jest*, to get together with God on the issue of death.